THE GRASSHOPPER:
GAMES, LIFE AND UTOPIA

DRAWINGS BY FRANK NEWFELD

BERNARD Herbert SUITS

The Grasshopper

GAMES, LIFE
AND UTOPIA

UNIVERSITY OF TORONTO PRESS

Toronto Buffalo

© University of Toronto Press 1978
Toronto Buffalo London
Printed in Canada

Design: William Rueter

Library of Congress Cataloguing in Publication Data

Suits, Bernard Herbert, 1925–
   The grasshopper

   Includes index
   1.   Utopias.   2.   game theory
   I.   Title
   HX806.S78   335'.02   78-8989
   ISBN 0-8020-2301-0

For Nancy and Mark and Conn

Contents

Preface

The Grasshopper of this book is the same Grasshopper whom Aesop made everlastingly famous as the model of improvidence. But while Aesop was content to cast this remarkable creature as the hero of a cautionary tale, he appears here as the exemplification – and articulate expositor – of the life most worth living. Because he is a working Utopian whose time has not yet come, he is destroyed by his uncompromising dedication to a premature ideal. But because he is also a speculative Utopian, he is able to defend that ideal – and the death which is the predictable consequence of its whole-hearted pursuit – before the end comes. Central to that defence is the Grasshopper's claim that Utopian existence is fundamentally concerned with game-playing, and so the book is largely devoted to formulating a theory of games.

That theory is not intended to be in any direct way a contribution to the field of investigation known officially as Game Theory, although it is possible that some game theorists may find it of more than marginal interest. Nor is the book essentially a contribution to sociology or social psychology, although it contains an extended discussion of role-playing and one section is addressed to Eric Berne's *Games People Play*.

The orientation of the book is philosophical in one traditional sense of that word. It is the attempt to discover and formulate a definition, and to follow the implications of that discovery even when they lead in surprising, and sometimes disconcerting, directions.

I am aware, of course, of the fairly widespread disenchantment with the search for definitions that currently prevails in the philosophical community, and indeed in the intellectual community generally. And Wittgenstein, one of the most forceful spokesmen (and certainly the most exotic) for the anti-definitional attitude, is famous for having singled out the attempt to define games as illustrating *par excellence* the futility of attempting to define anything whatever. 'Don't say,' Wittgenstein admonishes us, ' "there must be something common or they would not be called 'games' " – but *look and see* whether there is anything common to all.' This is unexceptionable advice. Unfortunately, Wittgenstein himself did not follow it. He looked, to be sure, but because he had decided beforehand that games are indefinable, his look was fleeting, and he saw very little. So I invite the reader to join me in a longer and more penetrating look at games, and to defer judgment as to whether all games have something in common pending completion of such an inspection.

In order to avoid possible misunderstanding, I add a disclaimer. The following inquiry is not, and should not be taken to be, a kind of anti-anti-definitional manifesto, nor should it be seen as depending for its cogency upon a commitment to the universal fruitfulness of definition construction. It seems altogether more reasonable to begin with the hypothesis that some things are definable and some are not, and that the only way to find out which are which is to follow Wittgenstein's excellent advice and *look and see*.

Acknowledgments

Some parts of this book have appeared in print elsewhere. Chapter One, the last part of Chapter Three, and Chapter Fifteen are revised versions of parts of two essays first published in *The Philosophy of Sport: A Collection of Original Essays* (Charles C Thomas 1973); the first part of Chapter Three originally appeared under the title 'What Is a Game?' in *Philosophy of Science* 1967; Chapter Seven appeared under its present title in *Philosophy of Science* 1969; and several paragraphs in Chapter Three are taken from 'Is Life a Game We Are Playing?' which was first printed in *Ethics* 1967 (Copyright 1967 by the University of Chicago). Grateful acknowledgment is made to the publishers for permission to include that material here.

This book has been published with the help of a grant from the Humanities Research Council of Canada, using funds provided by the Canada Council.

The book owes its existence in large part to a number of people for their encouragement of and interest in my pursuit of the study of games over the years. For that I thank Charner Perry, Richard Rudner, J. Sayer Minas, Nathan Brett, and especially Jan Narveson. I would like to acknowledge a genuinely delightful association with the staff of University of Toronto Press, especially R.I.K. Davidson, Jean Jamieson, Margaret Parker, and Laurie Lewis. Finally, I thank Frank Newfeld for the graphic wit he has brought to my text.

THE GRASSHOPPER:
GAMES, LIFE AND UTOPIA

The players

THE GRASSHOPPER	A shiftless but thoughtful practitioner of applied entomology
SKEPTICUS and PRUDENCE	Disciples of the Grasshopper
PROFESSOR SNOOZE	An accident-prone academic
DR THREAT	A murderer
SMITH and JONES	Two supporting players with a penchant for getting themselves into sticky but illustrative situations
ROBINSON	A friend of Smith and Jones who is invoked by them when needed
IVAN and ABDUL	Two retired army officers looking for a bit of fun
THE VOICE OF LOGIC	Nemesis of Ivan and Abdul
SIR EDMUND HILLARY	A mountain climber
PORPHYRYO SNEAK	The greatest spy in the world
BARTHOLOMEW DRAG	The greatest bore in the world
DR HEUSCHRECKE	A therapist consulted by Sneak and Drag
JOHN STRIVER and WILLIAM SEEKER	Two disgruntled utopians

...if there were no winters to guard against, then the Grasshopper would not get his comeuppance, nor the ant his shabby victory.

newfeld 78

1 Death of the Grasshopper

In which the Grasshopper, after defending to his disciples his way of life and impending death, dies

It was clear that the Grasshopper would not survive the winter, and his followers had gathered round him for what would no doubt be one of their last meetings. Most of them were reconciled to his approaching death, but a few were still outraged that such a thing could be allowed to happen. Prudence was one of the latter, and she approached the Grasshopper with a final plea. 'Grasshopper,' she said, 'a few of us have agreed to give up a share of our food to tide you over till spring. Then next summer you can work to pay us back.'

'My dear child,' responded the Grasshopper, 'you still don't understand. The fact is that I will *not* work to pay you back. I will not work at all. I made that perfectly clear, I thought, when the ant turned me away from his door. My going to him in the first place was, of course, a mistake. It was a weakness to which I shall not give in again.'

'But,' continued Prudence, 'we don't begrudge you a portion of our food. If you like, we will not require you to pay us back. We are not, after all, ants.'

'No,' replied the Grasshopper, 'you are not ants, not any more. But neither are you grasshoppers. Why should you give me the fruits of your labour? Surely that would not be just, when I tell you quite clearly that I will not pay you back.'

'But *that* kind of justice,' exclaimed Prudence, 'is only the justice of ants. Grasshoppers have nothing to do with such "justice." '

'You are right,' said the Grasshopper. 'The justice which is fairness in trading is irrelevant to the lives of true grasshoppers. But there is a different kind of justice which prevents me from accepting your offer. Why are you willing to work so that I may live? Is it not because I embody in my life what you aspire to, and you do not want the model of your aspirations to perish? Your wish is understandable, and to a certain point even commendable. But at bottom it is inconsistent and self-defeating. It is also – and I hope you will not take offence at my blunt language – hypocritical.'

'Those are hard words, Grasshopper.'

'But well meant. My life, you must understand, was not intended to be a sideshow, yet that seems to be what you want to make of it. You should value me because you want to be like me, and not merely so that you can boast to the ants that you are an intimate of the Grasshopper, that oddity of nature.'

'We have never done that, Grasshopper!'

'I believe you. But you might as well have done so if you believe that your proposal is a good one. For it amounts to working because I will not. But the whole burden of my teaching is that you ought to be idle. So now you propose to use me as a pretext not only for working, but for working harder than ever, since you would have not only yourselves to feed, but me as well. I call this hypocritical because you would like to take credit for doing something which is no more than a ruse for avoiding living up to your ideals.'

At this point Skepticus broke in with a laugh. 'What the Grasshopper means, Prudence,' he said, 'is that we do not yet have the courage of his convictions. The point is that we should not only refuse to work for the Grasshopper, we should also refuse to work for ourselves. We, like him, should be dying for our principles. That we are not is the respect in which, though no longer ants, we are not grasshoppers either. And, of course, given the premise that the life of the Grasshopper is the only life worth living, what he says certainly follows.'

'Not quite, Skepticus,' put in the Grasshopper. 'I agree that the principles in question are worth dying for. But I must remind you that they are the principles of Grasshoppers. I am not here to persuade you to die for my principles, but to persuade you that *I* must. We ought to be quite clear about our respective roles. You are not here to die for me, but I for you. You only need, as Skepticus put it, the courage of my convictions up to a point; that is, courage sufficient to approve rather than to deplore my death. Neither of you is quite prepared to grant that approval, though for different reasons. You, Prudence, because, although you believe the principles are worth dying for, you do not believe they need to be died for; and you, Skepticus, because you are not even sure that the principles are worth dying for.'

'Although,' replied Skepticus, 'I believe you to be the wisest being alive – which is why I have never left your side during the whole summer of your life – I have to admit that I am still not convinced that the life of the Grasshopper is the best life to live. Perhaps if you could give me a clearer vision of the good life as you see it my convictions would approach yours, and my courage as well. You might do this by one of the parables for which you are justly esteemed.'

'Parables, my dear Skepticus,' replied the Grasshopper, 'ought to come at the end, not at the beginning, of serious inquiry; that is, only at the point where arguments fail. But speaking of parables, you may be sure that the ants will fashion one out of my career. They will very likely represent my life as a moral tale, the point of which is the superiority of a prudent to an idle way of life. But it should really be the Grasshopper who is the hero of the tale; it is he, not the ant, who should have the hearer's sympathy. The point of the parable should be not the ant's triumph, but the Grasshopper's tragedy. For one cannot help reflecting that if there were no winters to guard against, then the Grasshopper would not get his come-uppance nor the ant his shabby victory. The life of the Grasshopper would be vindicated and that of the ant absurd.'

'But there *are* winters to guard against,' Prudence protested.

'No doubt. Still, it is possible that with accelerating advances in technology the time will come when there are in fact no winters. We may therefore conclude that although my timing may be a bit off, my way of life is not wrong in principle.'

'The operation was successful but the patient died,' put in Skepticus.

'No,' replied the Grasshopper, 'it's not quite like that. That my way of life may eventually be vindicated in practice is, now that I think of it, really beside the point. Rather, it is the *logic* of my position which is at issue. And this logic shows that prudential actions (e.g., those actions we ordinarily call work) are self-defeating in principle. For prudence may be defined as the disposition 1/ to sacrifice something good (e.g., leisure) if and only if such sacrifice is necessary for obtaining something better (e.g., survival), and 2/ to reduce the number of good things requiring sacrifice – ideally, at least – to zero. The ideal of prudence, therefore, like the ideal of preventive medicine, is its own extinction. For if it were the case that no sacrifices of goods needed ever to be made, then prudential actions would be pointless, indeed impossible. This principle, knowledge of which I regard as an indispensable first step on the path to wisdom, the ants seem never even to have entertained. The true Grasshopper sees that work is not self-justifying, and that his way of life is the final justification of any work whatever.'

'But surely,' replied Skepticus, 'you are carrying your point to an unreasonable extreme. You talk as though there were but two possible alternatives: either a life devoted exclusively to play or a life devoted exclusively to work. But most of us realize that our labour is valuable because it permits us to play, and we are presumably seeking to achieve some kind of balance between work input and play output. People are not, and do not want to be, wholly grasshoppers or wholly ants, but a combination of the two; people are and want to be (if you will forgive a regrettably vulgar but spooneristically inevitable construction) asshoppers or grants. We can, of course, all cease to work, but if we do then we cannot play for long either, for we will shortly die.'

'I have three answers to make to what you have said, Skepticus, and I fear I shall have to make them quickly, for the sun has set and the frost is already creeping through the fields. First, evidently I was put on earth just to play out my life and die, and it would be impious of me to go against my destiny. That is, if you like, the theology of the case. But second, there is also a logic of the case which is as inescapable as fate or, if you like, a fate of the case which is as inescapable as logic. The only argument against living the life of the Grasshopper arises from the contingent fact that at present one dies if one does not work. The answer to that argument is that my death is inevitable in any case. For if I am *improvident* in summer, then I will die in winter. And if I am *provident* in summer, then I will cease to be the Grasshopper, by definition. But I will be either provident or improvident in summer; there is no third alternative. Therefore, either I die or I cease to be the Grasshopper. But since I am just the Grasshopper, no more and no less, dying and ceasing to be the Grasshopper are one and the same thing for me. I cannot escape that logic or that fate. But since I am the Grasshopper and you are not, it would seem to follow that you are not compelled by this logic. As I intimated earlier, I often think that I was put on earth just to die for you; to bear that heavy but inevitable cross. But I confess that that is when I am in something of an early Christian – or late pagan – frame of mind. At other times (and this brings me to my third and final answer to your objection, Skepticus) I have the oddest notion that both of you are Grasshoppers in disguise; in fact, that everyone alive is really a Grasshopper.'

At this Prudence whispered to Skepticus, 'The end must be near; his mind is beginning to wander.' But Skepticus just looked keenly at their friend and teacher as he continued to speak.

'I admit that this is a wild fancy,' the Grasshopper was saying, 'and I hesitate to tell you my thoughts. Still, I am used to being thought foolish, so I shall proceed, inviting you to make of my words what you will. Then let me tell you that I have always had a recurring dream, in which it is revealed to me – though how it is revealed I cannot say – that everyone alive is in fact engaged in playing elaborate games, while at the same time believing themselves to be going about their ordinary affairs. Carpenters, believing themselves to be merely pursuing their trade, are really playing a game, and similarly with politicians, philosophers, lovers, murderers, thieves, and saints. Whatever occupation or activity you can think of, it is in reality a game. This revelation is, of course, astonishing. The sequel is terrifying. For in the dream I then go about persuading everyone I find of the great truth which has been revealed to me. How I am able to persuade them I do not know, though persuade them I do. But precisely at the point when each is persuaded – and this is the ghastly part – each ceases to exist. It is not just that my auditor vanishes on the spot, though indeed he does. It is that I also know with absolute certainty that he no longer exists anywhere. It is as though he had never been. Appalled as I am by the results of my teaching, I cannot stop, but quickly move on to the next creature with my news, until I have preached the truth throughout the universe and have converted everyone to oblivion. Finally I stand alone beneath the summer stars in absolute despair. Then I awaken to the joyful knowledge that the world is still teeming with sentient beings after all, and that it was only a dream. I see the carpenter and philosopher going about their work as before ... But is it, I ask myself, just as before? Is the carpenter on his roof-top simply hammering nails, or is he making some move in an ancient game whose rules he has forgotten! But now the chill creeps up my legs. I grow drowsy. Dear friends, farewell.'

Oh, Skepticus, how maddening
I thought I had finally
come to understand the
message of

newfeld '78

2 Disciples

In which Skepticus and Prudence discover that the Grasshopper has left them with a tangle of riddles about play, games, and the good life

The next day Skepticus called upon a grieving Prudence.

SKEPTICUS: It is time to put aside your grief, my girl, and help me examine our bequest.

PRUDENCE: (*drying her eyes*) What bequest?

S: Why, the Grasshopper's dream, of course. I have been awake the whole night trying to puzzle it out.

P: (*blowing her nose*) It was certainly very strange.

S: Yes, it was. But it strikes me that even stranger than the dream itself was the Grasshopper's telling us about it at all.

P: Why do you say that?

S: Well, the Grasshopper didn't tell us about the dream just because it was an interesting dream. He brought it up in the course of answering a question I had asked him. In effect, I had put it to him that while all work and no play undoubtedly makes Jack a dull ant, all play and no work makes Jack a dead grasshopper.

P: Yes, you were challenging him to justify his existence.

S: Quite so. And he made three replies to that challenge. The first he called the theological answer and the second he called the logical answer.

P: That's right.

S: And what about the third answer, Prudence?

P: The third answer was the dream.

S: Yes, a dream about people playing games. That is what is so strange.

P: What is so strange about that? Surely the strangeness lies in the fact that they were playing *unconscious* games, and that they vanished as soon as they realized that that was what they were doing.

s: Oh, that is strange, I grant you. But it is just the kind of strangeness you expect a dream to have. There is, however, another and, so to say, *prior* strangeness which must be fathomed before we can begin to fathom the strangeness of the dream itself.

p: What on earth are you talking about, Skepticus?

s: I am saying that there is a question we have to answer before we can solve the puzzle of the dream.

p: What question?

s: This question: Why were the creatures in the Grasshopper's dream playing games instead of the trombone?

p: Skepticus, I haven't the foggiest idea what you're talking about.

s: I am trying to get at the point of the Grasshopper's third answer, Prudence. His first two answers – the theological answer and the logical answer – really amounted to the same thing, did they not? Each was a way of expressing the Grasshopper's determination to remain true to himself, even at the cost of his life.

p: Yes, that's right.

s: And his remaining true to himself, Prudence, what did that consist in?

p: Why, in refusing to work and insisting upon devoting himself exclusively to play.

s: And what did the words 'work' and 'play' mean in that context?

p: Pretty much what most people usually mean by those words, I should think. Working is doing things you have to do and playing is doing things for the fun of it.

s: So that for 'play' we could substitute the expression 'doing things we value for their own sake,' and for 'work' we could substitute the expression 'doing things we value for the sake of something else.'

p: Yes. Work is a kind of necessary evil which we accept because it makes it possible for us to do things we think of as being good in themselves.

s: So that under the heading *play* we could include any number of quite different things: vacationing in Florida, collecting stamps, reading a novel, playing chess, or playing the trombone?

p: Yes, all of those things would count as 'play' as we are using the word. We are using 'play' as equivalent to 'leisure activities.'

s: Then it is clear, is it not, that 'playing,' in this usage, cannot be the same as 'playing games,' since there are many leisure activities, as we have just noted, that are not games.

p: No, they are not the same; playing games is just one kind of leisure activity.

s: Therefore, when the Grasshopper was extolling the life of play he meant by that life, presumably, not doing any specific thing, but doing any of a number of quite different things, depending, no doubt, on the talents and preferences of those doing the playing. That is, some people like to collect stamps, and some do not. Some have a talent for chess or for playing wind instruments, and some do not. So the Grasshopper surely was not arguing that the life he was seeking to justify – the life of the Grasshopper – was identical with just *one* of these leisure activities. He was not contending, for example, that the life of the Grasshopper is identical with playing the trombone.

p: Of course not, Skepticus, how absurd!

s: Yes, that would be absurd. And that is precisely why I find the Grasshopper's third answer so strange. For in that answer he seemed to be taking the view not that the life of the Grasshopper ought to consist simply in leisure activities, but that it ought to consist in playing *games*. For he began his answer, you will recall, by telling us that he sometimes fancied that everyone alive was really a grasshopper in disguise.

p: Yes, I remember.

s: And then, presumably as an explanation of what he meant by that curious observation, he began to tell us about his dream, in which everyone alive was playing games but did not know that they were playing games. The conclusion seems inescapable that the Grasshopper was thinking of a grasshopper in disguise as being identical with someone playing a game without knowing that he was playing a game, and that he therefore believed *game* playing, and not merely playing in general, to be the essential life of the grasshopper.

p: Yes, I see, Skepticus. How very odd.

s: Indeed. For the dream is revealed as a riddle which is itself contained within another riddle. First there is the rather complicated riddle of the dream itself. Why should creatures who do not know themselves to be grasshoppers, and who have been playing games thay they do not know to be games, suffer annihilation upon discovering that that is what they have been doing; and why, if they are playing games, don't they know it? But all of this is part of another riddle. That is, why should the quintessential grasshopper be a player of games rather than a doer of any number of others things which are valuable in themselves and which therefore count as 'play' every bit as much as game playing does?

p: Oh, Skepticus, how maddening! I thought I had finally come to understand the message of the Grasshopper. But now it appears that his most profound teaching will be for ever lost to us.

s: Not necessarily, Prudence.

p: What do you mean?

s: Perhaps the Grasshopper will be resurrected.

p: Resurrected!

s: Well, he seemed to regard himself as a combination of Socrates and Jesus Christ.

p: Skepticus!

s: Still, I don't think I'll wait for that much-to-be-hoped-for development.

p: Do you think you can solve the riddles by yourself?

s: At any rate, I propose to try. You heard me mention, when we were talking to the Grasshopper, that I had never left his side all summer long?

p: Yes.

s: Well, what do you suppose we talked about from May till September?

p: The Grasshopper's philosophy of life, I suppose.

s: More particularly, Prudence, we talked about *games*.

p: Games! Then you weren't really surprised, were you, when the Grasshopper told us his dream of game players?

s: Perhaps I should not have been, Prudence, but I was. You see, I had simply assumed, without much thinking about it, that the Grasshopper was interested in talking about games because he happened to be more interested in games than in some other playtime pursuit that we might just as well have discussed.

p: Like music if the Grasshopper had been, say, a trombone player.

s: Precisely. Of course, now I see that there was a good deal more to it than that.

p: Well, tell me, Skepticus. What did the Grasshopper say about games?

s: First he presented a definition of games or, to be more precise, a definition of game playing. Then he invited me to subject that definition to a series of tests. I was to advance against the definition the most compelling objections I could devise, and he was to answer those objections.

p: And did the definition withstand your attacks?

s: He was able, or so it seemed to me, to defend the definition against all of my challenges. Furthermore, in the course of meeting those challenges a number of features of game playing not contained in the definition itself were brought to light, so that at the end we had developed a rather elaborated outline, at least, of a general theory of games. Fortunately I took careful notes of those conversations, and so I propose to reconstruct the argument just as it evolved. For I am convinced that the solution of the complicated riddle which the Grasshopper has bequeathed to us lies in the nature of games. And I am sure, now, that the Grasshopper spoke to us in a dream parable – instead of telling us straight out what he had in mind – precisely because he had spent the whole summer providing me with all the clues necessary for solving that riddle.

p: Why, Skepticus, it is almost as though he was –

s: Playing a game with us?

p: So it would appear, Skepticus. Begin your reconstruction at once, then, so that the game can begin.

s: Very well, Prudence, if it is quite fitting to call a game an enterprise which aims at nothing less than an elucidation of Grasshopper logic, an examination of Grasshopper ideals, and an interpretation of Grasshopper dreams.*

---

* I have divided the Grasshopper's discourse on games into chapters and, in some cases, into chapter sub-sections, and I have added my own titles and sub-titles to these divisions. I am also responsible for footnoting the Grasshopper's citation of other sources on the subject of games (save for Chapter Seven, where the Grasshopper has provided his own notes), but in all other respects what follows is a faithful account of our inquiry just as it progressed – Skepticus.

It is impossible to win a game and at the same time to

newfeld '78

3 Construction of a definition

The beginning of a flashback which continues to Chapter Thirteen. Here the Grasshopper arrives at a definition of games by two different routes

Game playing as the selection of inefficient means

Mindful of the ancient canon that the quest for knowledge obliges us to proceed from what is more obvious to what is less obvious [began the Grasshopper], let us start with the commonplace belief that playing games is different from working. Games therefore might be expected to be what work, in some salient respect, is not. Let us now baldly characterize work as 'technical activity,' by which I mean activity in which an agent seeks to employ the most efficient available means for reaching a desired goal. Since games, too, evidently have goals, and since means are evidently employed for their attainment, the possibility suggests itself that games differ from technical activities in that the means employed in games are not the most efficient. Let us say, then, that games are goal-directed activities in which inefficient means are intentionally chosen. For example, in racing games one voluntarily goes all round the track in an effort to arrive at the finish line instead of 'sensibly' cutting straight across the infield.

The following considerations, however, seem to cast doubt on this proposal. The goal of a game, we may say, is winning the game. Let us take an example. In poker I am a winner if I have more money when I stop playing than I had when I started. But suppose that one of the other players, in the course of the game, repays me a debt of a hundred dollars, or suppose I hit another player on the head and take all of his money from him. Then, although I have not won a single hand all evening, am I nevertheless a winner? Clearly not, since I did not increase my money as a consequence of playing poker. In order to be a winner (a sign and product of which is, to be sure, the gaining of money) certain conditions must be met which are not met by the collection of a debt or by felonious assault. These conditions are the rules of poker, which tell us what we can and what we cannot do with the cards and the money. Winning at poker consists in increasing one's money by using only means permitted by the rules, although mere obedience to the rules does not by itself ensure victory. Better and worse means are equally permitted by the rules. Thus in Draw Poker retaining an ace along with a pair and discarding the ace while retaining the pair are both permissible plays, although one is usually a better play than the other. The means for winning at poker, therefore, are limited, but not completely determined, by the rules. Attempting to win at poker may accordingly be described as attempting to gain money by using the most efficient means available, where only those means permitted by the rules are available. But if that is so, then playing poker is a technical activity as originally defined.

Still, this seems a strange conclusion. The belief that working and playing games are quite different things is very widespread, yet we seem obliged to say that playing a game is just another job to be done as competently as possible. Before giving up the thesis that playing a game involves a sacrifice of efficiency, therefore, let us consider one more example. Suppose I make it my purpose to get a small round object into a hole in the ground as efficiently as possible. Placing it in the hole with my hand would be a natural means to adopt. But surely I would not take a stick with a piece of metal on one end of it, walk three or four hundred yards away from the hole, and then attempt to propel the ball into the hole with the stick. That would not be technically intelligent. But such an undertaking is an extremely popular game, and the foregoing way of describing it evidently shows how games differ from technical activities.

But of course it shows nothing of the kind. The end in golf is not correctly described as getting a ball into a hole in the ground, or even, to be more precise, into several holes in a set order. It is to achieve that end with the smallest possible number of strokes. But a stroke is a certain type of swing with a golf club. Thus, if my end were simply to get a ball into a number of holes in the ground, I would not be likely to use a golf club in order to achieve it, nor would I stand at a considerable distance from each hole. But if my end were to get a ball into some holes with a golf club while standing at a considerable distance from each hole, why then I would certainly use a golf club and I would certainly take up such positions. Once committed to that end, moreover, I would strive to accomplish it as efficiently as possible. Surely no one would want to maintain that if I conducted myself with utter efficiency in pursuit of this end I would not be playing a game, but that I *would* be playing a game just to the extent that I permitted my efforts to become sloppy. Nor is it the case that my use of a golf club is a less efficient way to achieve my end than would be the use of my hand. To refrain from using a golf club as a means for sinking a ball with a golf club is not more efficient because it is not possible. Inefficient selection of means, accordingly, does not seem to be a satisfactory account of game playing.

The inseparability of rules and ends in games

The objection advanced against the last thesis rests upon, and thus brings to light, consideration of the place of rules in games: they seem to stand in a peculiar relation to ends. The end in poker is not simply to gain money, or in golf simply to get a ball into a hole, but to do these things in prescribed (or, perhaps more accurately, not to do them in proscribed) ways; that is, to do them only in accordance with rules. Rules in games thus seem to be in some sense inseparable from ends, for to break a game rule is to render impossible the attainment of an end. Thus, although you may receive the trophy by lying about your golf score, you have certainly not won the game. But in what we have called technical activity it *is* possible to gain an end by breaking a rule; for example, gaining a trophy by lying about your golf score. So while it is possible in a technical action to break a rule without destroying the original end of the action, in games the reverse appears to be the case. If the rules are broken the original end becomes impossible of attainment, since one cannot (really) win the game unless one plays it, and one cannot (really) play the game unless one obeys the rules of the game.

This may be illustrated by the following case. Professor Snooze has fallen asleep in the shade provided by some shrubbery in a secluded part of the campus. From a nearby walk I observe this. I also notice that the shrub under which he is reclining is a man-eating plant, and I judge from its behaviour that it is about to eat the man Snooze. As I run across to him I see a sign which reads KEEP OFF THE GRASS. Without a qualm I ignore this prohibition and save Snooze's life. Why did I make this (no doubt scarcely conscious) decision? Because the value of saving Snooze's life (or of saving a life) outweighed the value of obeying the prohibition against walking on the grass.

Now the choices in a game appear to be radically unlike this choice. In a game I cannot disjoin the end, winning, from the rules in terms of which winning possesses its meaning. I can, of course, decide to cheat in order to gain the pot, but then I have changed my end from winning a game to gaining money. Thus, in deciding to save Snooze's life my purpose was not 'to save Snooze while at the same time obeying the campus rules for pedestrians.' My purpose was to save Snooze's life, and there were alternative ways in which this might have been accomplished. I could, for example, have remained on the sidewalk and shouted to Snooze in an effort to awaken him. But precious minutes might have been lost, and in any case Snooze, although he tries to hide it, is nearly stone deaf. There are evidently two distinct ends at issue in the Snooze episode: saving Snooze and obeying the rule, out of respect either for the law or for the lawn. And I can achieve either of these ends without at the same time achieving the other. But in a game the end and the rules do not admit of such disjunction. It is impossible for me to win the game and at the same time to break one of its rules. I do not have open to me the alternatives of winning the game honestly and winning the game by cheating, since in the latter case I would not be playing the game at all and thus could not, *a fortiori*, win it.

Now if the Snooze episode is treated as an action which has one, and only one, end – (Saving Snooze) *and* (Keeping off the grass) – it can be argued that the action has become, just by virtue of that fact, a game. Since there would be no independent alternatives, there would be no choice to be made; to achieve one part of the end without achieving the other part would be to fail utterly. On such an interpretation of the episode suppose I am congratulated by a grateful faculty for my timely intervention. A perfectly appropriate response would be: 'I don't deserve your praise. True, I saved Snooze, but since I walked on the grass it doesn't count,' just as though I were to admit to carrying the ball to the cup on the fifth green. Or again, on this interpretation, I would originally have conceived the problem in a quite different way: 'Let me see if I can save Snooze without walking on the grass.' One can then imagine my running as fast as I can (but taking no illegal short cuts) to the Athletic Building, where I request (and meticulously sign out) a pole vaulter's pole with which I hope legally to prod Snooze into wakefulness, whereupon I hurry back to Snooze to find him disappearing into the plant. 'Well,' I remark, not without complacency, 'I didn't win, but at least I played the game.'

It must be pointed out, however, that this example could be misleading. Saving a life and keeping off the grass are, as values, hardly on the same footing. It is possible that the Snooze episode appears to support the contention at issue (that games differ from technical actions because of the inseparability of rules and ends in the former) only because of the relative triviality of one of the alternatives. This peculiarity of the example can be corrected by supposing that when I decide to obey the rule to keep off the grass, my reason for doing so is that I am a kind of demented Kantian and thus regard myself to be bound by the most weighty philosophical considerations to honour *all* laws with equal respect. So regarded, my maddeningly proper efforts to save a life would not appear ludicrous but would constitute moral drama of the highest order. But since we are not demented Kantians, Skepticus, a less fanciful though logically identical example may be cited.

Let us suppose the life of Snooze to be threatened not by a man-eating plant but by Dr Threat, who is found approaching the snoozing Snooze with the obvious intention of murdering him. Again I want to save Snooze's life, but I cannot do so (let us say) without killing Threat. However, there is a rule to which I am very strongly committed which forbids me to take another human life. Thus, although (as it happens) I could easily kill Threat from where I stand (with a loaded and cocked pistol I happen to have in my hand), I decided to try to save Snooze by other means, just because of my wish to obey the rule which forbids killing. I therefore run towards Threat with the intention of wresting the weapon from his hand. I am too late, and he murders Snooze. This seems to be a clear case of an action having a conjunctive end of the kind under consideration, but one which we are not at all inclined to call a game. My end, that is to say, was not simply to save the life of Snooze, just as in golf it is not simply to get the ball into the hole, but to save his life without breaking a certain rule. I want to put the ball into the hole fairly and I want to save Snooze morally. Moral rules are perhaps generally regarded as figuring in human conduct in just this fashion. Morality says that if something can be done only immorally it ought not to be done at all. 'What profiteth it a man,' etc. The inseparability of rules and ends does not, therefore, seem to be a completely distinctive characteristic of games.

Game rules as not ultimately binding

It should be noticed, however, that the foregoing criticism requires only a partial rejection of the proposal at issue. Even though the attack seems to show that not all things which correspond to the formula are games, it may still be the case that all games correspond to the formula. This suggests that we ought not to reject the proposal but ought first to try to limit its scope by adding to it an adequate differentiating principle. Such a principle is suggested by the striking difference between the two Snooze episodes that we have noted. The efforts to save Snooze from the man-eating plant without walking on the grass appeared to be a game because saving the grass strikes us as a trifling consideration when compared with saving a life. But in the second episode, where KEEP OFF THE GRASS is replaced by THOU SHALT NOT KILL, the situation is quite different. The difference may be put in the following way. The rule to keep off the grass is not an ultimate command, but the rule to refrain from killing perhaps is. This suggests that, in addition to being the kind of activity in which rules are inseparable from ends, games are also the kind of activity in which commitment to these rules is never ultimate. For the person playing the game there is always the possibility of there being a non-game rule to which the game rule may be subordinated. The second Snooze episode is not a game, therefore, because the rule to which the rescuer adheres, even to the extent of sacrificing Snooze for its sake, is, for him, an ultimate rule. Rules are always lines that we draw, but in games the lines are always drawn short of a final end or a paramount command. Let us say, then, that a game is an activity in which observance of rules is part of the end of the activity, and where such rules are non-ultimate; that is, where other rules can always supersede the game rules; that is, where the player can always stop playing the game.

However, consider the Case of the Dedicated Driver. Mario Stewart (the driver in question) is a favoured entrant in the motor car race of the century at Malaise. And in the Malaise race there is a rule which forbids a vehicle to leave the track on pain of disqualification. At a crucial point in the race a child crawls out upon the track directly in the path of Mario's car. The only way to avoid running over the child is to leave the track and suffer disqualification. Mario runs over the child and completes the race. I submit that we ought not, for this reason, to deny that he is playing a game. It no doubt strikes us as inappropriate to say that a person who would do such a thing is (merely) playing. But the point is that Mario is not playing in an unqualified sense, he is playing a *game*. And he is evidently playing it more whole-heartedly than the ordinary driver is prepared to play it. From his point of view a racer who turned aside instead of running over the child would have been playing *at* racing; that is, he would not have been a dedicated player. But it would be paradoxical indeed if supreme dedication to an activity somehow vitiated the activity. We do not say that a man is not really digging a ditch just because his whole heart is in it.

However, the rejoinder may be made that, to the contrary, that is just the mark of a game: it, unlike digging ditches, is just the kind of thing which cannot command ultimate loyalty. That, it may be contended, is precisely the force of the proposal about games under consideration. And in support of this contention it might be pointed out that it is generally acknowledged that games are in some sense non-serious undertakings. We must therefore ask in what sense games are, and in what sense they are not, serious. What is believed when it is believed that games are not serious? Not, certainly, that the players of games always take a very light-hearted view of what they are doing. A bridge player who played his cards randomly might justly be accused of *failing* to take the game seriously – indeed, of failing to play the game at all just because of his failure to take it seriously. It is much more likely that the belief that games are not serious means what the proposal under consideration implies: that there is always something in the life of a player of a game more important than playing the game, or that a game is the kind of thing that a player could always have reason to stop playing. It is this belief which I would like to question.

Let us consider a golfer, George, so devoted to golf that its pursuit has led him to neglect, to the point of destitution, his wife and six children. Furthermore, although George is aware of the consequences of his mania, he does not regard his family's plight as a good reason for changing his conduct. An advocate of the view that games are *not* serious might submit George's case as evidence for that view. Since George evidently regards nothing in his life to be more important than golf, golf has, for George, *ceased to be a game*. And this argument would seem to be supported by the complaint of George's wife that golf is for George no longer a game, but a way of life.

But we need not permit George's wife's observation to go un-challenged. The correctness of saying that for George golf is no longer merely a form of recreation may be granted. But to argue that George's golf playing is for that reason not a game is to assume the very point at issue, which is whether a game can be of supreme importance to anyone. Golf, to be sure, is taking over the whole of George's life. But it is, after all, the game which is taking over his life, and not something else. Indeed, if it were not a game which had led George to neglect his duties, his wife might not be nearly as outraged as she is; if, for example, it had been good works, or the attempt to formulate a definition of game playing. She would no doubt still deplore such extra-domestic pre-occupation, but to be kept in rags because of a game must strike her as an altogether different order of deprivation.

Supreme dedication to a game, as in the cases of the auto racer and George, may be repugnant to nearly everyone's moral sense. That may be granted – indeed, insisted upon, since our loathing is excited by the very fact that it is a game which has usurped the place of ends we regard as so much more worthy of pursuit. Thus, although such behaviour may tell us a good deal about such players of games, I submit that it tells us nothing about the games they play. I believe that these observations are sufficient to discredit the thesis that game rules cannot be the object of an ultimate, or unqualified, commitment.

Means, rather than rules, as non-ultimate

I want to agree, however, with the general contention that in games there is something which is significantly non-ultimate, that there is a crucial limitation. But I would like to suggest that it is not the rules which suffer such limitation. Non-ultimacy evidently attaches to games at a quite different point. It is not that obedience to game rules must fall short of ultimate commitments, but that the means which the rules permit must fall short of ultimate utilities. If a high-jumper, for example, failed to complete his jump because he saw that the bar was located at the edge of a precipice, this would no doubt show that jumping over the bar was not the overriding interest of his life. But it would not be his refusal to jump to his death which would reveal his conduct to be a game; it would be his refusal to use something like a ladder or a catapult in his attempt to clear the bar. The same is true of the dedicated auto racer. A readiness to lose the race rather than kill a child is not what makes the race a game; it is the refusal to, *inter alia*, cut across the infield in order to get ahead of the other contestants. There is, therefore, a sense in which games may be said to be non-serious. One could intelligibly say of the high-jumper who rejects ladders and catapults that he is not serious about getting to the other side of the barrier. But one would also want to point out that he could be deadly serious about getting to the other side of the barrier *without* such aids, that is, about high-jumping. But whether games as such are less serious than other things would seem to be a question which cannot be answered solely by an investigation of games.

Consider a third variant of Snooze's death. In the face of Threat's threat to murder Snooze, I come to the following decision. I choose to limit myself to non-lethal means in order to save Snooze even though lethal means are available to me and I do not regard myself to be bound by any rule which forbids killing. (In the auto racing example the infield would *not* be filled with land mines.) And I make this decision even though it may turn out that the proscribed means are necessary to save Snooze. I thus make my end not simply saving Snooze's life, but saving Snooze's life without killing Threat, even though there appears to be no reason for restricting myself in this way.

One might then ask how such behaviour can be accounted for. And one answer might be that it is unaccountable, that it is simply arbitrary. However, the decision to draw an arbitrary line with respect to permissible means need not itself be an arbitrary decision. The decision to be arbitrary may have a purpose, and the purpose may be to play a game. It seems to be the case that the lines drawn in games are not really arbitrary at all. For both *that* the lines are drawn and also *where* they are drawn have important consequences not only for the type, but also for the quality, of the game to be played. It might be said that drawing such lines skilfully (and therefore not arbitrarily) is the very essence of the gamewright's craft. The gamewright must avoid two extremes. If he draws his lines too loosely the game will be dull because winning will be too easy. As looseness is increased to the point of utter laxity the game simply falls apart, since there are then no rules proscribing available means. (For example, a homing propellant device could be devised which would ensure a golfer a hole in one every time he played.) On the other hand, rules are lines that can be drawn too tightly, so that the game becomes too difficult. And if a line is drawn very tightly indeed the game is squeezed out of existence. (Suppose a game in which the goal is to cross a finish line. One of the rules requires the contestants to stay on the track, while another rule requires that the finish line be located in such a position that it is impossible to cross it without leaving the track.) The present proposal, therefore, is that games are activities in which rules are inseparable from ends (in the sense agreed to earlier), but with the added qualification that the means permitted by the rules are narrower in range than they would be in the absence of the rules.

Rules are accepted for the sake of the activity they make possible

Still, even if it is true that the function of rules in games is to restrict the permissible means to an end, it does not seem that this is by itself sufficient to exclude things which are not games. When I failed in my attempt to save Snooze's life because of my unwillingness to commit the immoral act of taking a life, the rule against killing functioned to restrict the means I would employ in my efforts to reach a desired end. What, then, distinguishes the cases of the high-jumper and auto racer from my efforts to save Snooze morally, or the efforts of a politician to get elected without lying? The answer lies in the reasons for obeying rules in the two types of case. In games I obey the rules just because such obedience is a necessary condition for my engaging in the activity such obedience makes possible. In high-jumping, as we have noted, although the contestants strive to be on the other side of a barrier, they voluntarily rule out certain means for achieving this goal. They will not walk around it, or duck under it, or use a ladder or catapult to get over it. The goal of the contestants is not to be on the other side of the barrier *per se*, since aside from the game they are playing they are unlikely to have any reason whatever for being on the other side. Their goal is not *simply* to get to the other side, but to do so only by using means permitted by rules, namely, by running from a certain distance and then jumping. And their *reason* for accepting such rules is just because they want to act within the limitations the rules impose. They accept rules so that they can play a game, and they accept these rules so that they can play this game.

But with respect to other rules – for example, moral rules – there is always another reason – what might be called an external or independent reason – for obeying whatever rule may be at issue. In behaving morally, we deny ourselves the option of killing a Threat or lying to the voters not because such denial provides us, like a high-jumper's bar, with an activity we would not otherwise have available to us, but because, quite aside from such considerations, we judge killing and lying to be wrong. The honest politician is not honest because he is interested primarily in the activity trying-to-get-elected-without-lying (as though he valued his commitment to honesty because it provided him with an interesting challenge), but for quite different reasons. He may, for example, be a Kantian, who believes that it is wrong, under any circumstances whatever, to lie. And so, since his morality requires him to be truthful in all cases, it requires him to be truthful in this case. Or he may be a moral teleologist, who believes that the consequences of dishonesty (either in this case or in general) work against practical possibilities which are in the long run more desirable than the possibility of being elected to office. But the high-jumper does not accept rules for either of these kinds of reason. He does not on principle always make things harder for himself; he does not even on principle always make surmounting physical barriers harder for himself. He does these things only when he wants to be engaged in high-jumping. Nor does the high-jumper, *qua* high-jumper, deny himself the use of more efficient means for clearing the bar because of higher priority moral claims (the catapult is being used to defend the town just now, or the ladder is being used to rescue a child from a rooftop), but just because, again, he wants to be high-jumping. In morals obedience to rules makes the action right, but in games it makes the action.

Of course it is not moral rules alone which differ from game rules in this respect. More generally, we may contrast the way that rules function in games with two other ways that rules function. 1/ Rules can be directives useful in seeking a given end (If you want to improve your drive, keep your eye on the ball), or 2/ they can be externally imposed limitations on the means that may be chosen in seeking an end (Do not lie to the public in order to get them to vote for you). In the latter way a moral rule, as we have seen, often functions as a limiting condition upon a technical activity, although a supervening technical activity can produce the same kind of limitation (If you want to get to the airport in time, drive fast, but if you want to arrive safely, don't drive too fast). Consider a ruled sheet of paper. I conform to these rules, when writing, in order to write straight. This illustrates the first kind of rule. Now suppose that the rules are not lines on a piece of paper, but paper walls which form a labyrinth, and while I wish to be out of the labyrinth I do not wish to damage the walls. The walls are limiting conditions on my coming to be outside. This illustrates the second kind of rule. 3/ Now returning to games, consider a third case. Again I am in the labyrinth, but my purpose is not just to *be* outside (as it might be if Ariadne were waiting for me to emerge), but to *get* out of the labyrinth, so to speak, labyrinthically. What is the status of the walls? It is clear that they are not simply impediments to my being outside the labyrinth, because it is not my purpose to (simply) be outside. For if a friend suddenly appeared overhead in a helicopter I would decline the offer of a lift, although I would accept it in the second case. My purpose is to get out of the labyrinth only by accepting the conditions it imposes, that is, by responding to the challenge it presents. Nor, of course, is this like the first case. There I was not interested in *seeing whether* I could write a sentence without breaking a rule, but in using the rules so that I could write straight.

We may therefore say that games require obedience to rules which limit the permissible means to a sought end, and where such rules are obeyed just so that such activity can occur.

Winning is not the end with respect to which rules limit means

There is, however, a final difficulty. To describe rules as operating more or less permissively with respect to means seems to conform to the ways in which we *invent* or *revise* games. But it does not seem to make sense at all to say that in games there are always means available for attaining one's end over and above the means permitted by the rules. Consider chess. The end sought by chess players, it would seem, is to win, which involves getting chess pieces onto certain squares in accordance with the rules of chess. But since to break a rule is to fail to attain that end, what other means are available? It was for just this reason that our very first proposal about the nature of games was rejected: using a golf club in order to play golf is not a less efficient, and therefore an alternative, means for seeking the end in question. It is a logically indispensable means.

The objection can be met, I believe, by pointing out that there is an end in chess analytically distinct from winning. Let us begin again, therefore, from a somewhat different point of view and say that the end in chess is, in a very restricted sense, to place your pieces on the board in such an arrangement that the opponent's king is, in terms of the rules of chess, immobilized. Now, without going outside chess we may say that the means for bringing about this state of affairs consist in moving the chess pieces. The rules of chess, of course, state how the pieces may be moved; they distinguish between legal and illegal moves. Since the knight, for example, is permitted to move in only a highly restricted manner, it is clear that the permitted means for moving the knight are of less scope than the possible means for moving him. It should not be objected at this point that other means for moving the knight – e.g., along the diagonals – are not really possible on the grounds that such use of the knight would break a rule and thus not be a means to winning. For the present point is not that such use of the knight would be a means to winning, but that it would be a possible (though not permissible) way in which to move the knight so that he would, for example, come to occupy a square so that, according to the rules of chess, the king would be immobilized. A person who made such a move would not, of course, be playing chess. Perhaps he would he cheating at chess. By the same token I would not be playing a game if I abandoned my arbitrary decision not to kill Threat while at the same time attempting to save Snooze. Chess and my third effort to save Snooze's life are games because of an 'arbitrary' restriction of means permitted in pursuit of an end.

The main point is that the end here in question is not the end of winning the game. There must be an end which is distinct from winning because it is the restriction of means to this other end which makes winning possible and also defines, in any given game, what it means to win. In defining a game we shall therefore have to take into account these two ends and, as we shall see in a moment, a third end as well. First there is the end which consists simply in a certain state of affairs: a juxtaposition of pieces on a board, saving a friend's life, crossing a finish line. Then, when a restriction of means for attaining this end is made with the introduction of rules, we have a second end, winning. Finally, with the stipulation of what it means to win, a third end emerges: the activity of trying to win – that is, playing the game.

And so when at the outset we entertained the possibility that games involved the selection of inefficient means, we were quite right. It is just that we looked for such inefficiency in the wrong place. Games do not require us to operate inefficiently with respect to *winning*, to be sure. But they do require us to operate inefficiently in trying to achieve that state of affairs which counts as winning only when it is accomplished according to the rules of the game. For the way in which those rules function is to prohibit use of the most efficient means for achieving that state of affairs.

The definition

My conclusion is that to play a game is to engage in activity directed towards bringing about a specific state of affairs, using only means permitted by rules, where the rules prohibit more efficient in favour of less efficient means, and where such rules are accepted just because they make possible such activity.

'Well, Skepticus,' concluded the Grasshopper, 'what do you think?'

'I think,' I replied, 'that you have produced a definition which is quite plausible.'

'But untested. I shall therefore ask you, Skepticus, to bend all of your considerable sceptical efforts to discrediting the definition. For if the definition can withstand the barrage of objections I believe I can count upon you to launch against it, then perhaps we shall be justified in concluding that the account is not merely plausible, but substantially correct. Will you help me with that task?'

'Gladly, Grasshopper,' I replied, 'if you will give me a moment to collect myself. For I feel as if we, too, had just succeeded in finding our way out of a complicated maze. I know that we have finally got clear, but I am quite unable to say how we managed to do it, for our correct moves are hopelessly confused in my mind with the false starts and blind alleys which formed so large a part of our journey. Just trying to think back over the twists and turns of the arguments makes me quite light-headed.'

'What you are describing, Skepticus, is a chronic but minor ailment of philosophers. It is called dialectical vertigo, and its cure is the immediate application of straightforward argumentation. In terms of your metaphor, you need to be suspended, as it were, over the maze, so that you can discriminate at a glance the true path from the false turnings. Let me try to give you such an overview of the argument.'

'By all means,' I said.

A more direct approach to games [continued the Grasshopper] can be made by identifying what might be called the *elements* of game-playing. Since games are goal-directed activities which involve choice, ends and means are two of the elements of games. But in addition to being means-end-oriented activities, games are also rule-governed activities, so that rules are a third element. And since, as we shall see, the rules of games make up a rather special kind of rule, it will be necessary to take account of one more element, namely, the attitudes of game players *qua* game players. I add '*qua* game players' because I do not mean what might happen to be the attitude of this or that game player under these or those conditions (e.g., the hope of winning a cash prize or the satisfaction of exhibiting physical prowess to an admiring audience), but the attitude without which it is not possible to play a game. Let us call this attitude, of which more presently, the *lusory* (from the Latin *ludus*, game) attitude.

My task will be to persuade you that what I have called the lusory attitude is the element which unifies the other elements into a single formula which successfully states the necessary and sufficient conditions for any activity to be an instance of game playing. I propose, then, that the elements of game are 1/ the goal, 2/ the means of achieving the goal, 3/ the rules, and 4/ the lusory attitude. I shall briefly discuss each of these in order.

The Goal We should notice first of all that there are three distinguishable goals involved in game playing. Thus, if we were to ask a long-distance runner his purpose in entering a race, he might say any one or all of three things, each of which would be accurate, appropriate, and consistent with the other two. He might reply 1/ that his purpose is to participate in a long-distance race, or 2/ that his purpose is to win the race, or 3/ that his purpose is to cross the finish line ahead of the other contestants. It should be noted that these responses are not merely three different formulations of one and the same purpose. Thus, winning a race is not the same thing as crossing a finish line ahead of the other contestants, since it is possible to do the latter unfairly by, for example, cutting across the infield. Nor is participating in the race the same as either of these, since the contestant, while fully participating, may simply fail to cross the finish line first, either by fair means or foul. That there must be this triplet of goals in games will be accounted for by the way in which lusory attitude is related to rules and means. For the moment, however, it will be desirable to select just one of the three kinds of goal for consideration, namely, the kind illustrated in the present example by crossing the finish line ahead of the other contestants. This goal is literally the simplest of the three, since each of the others presupposes it, while it does not presuppose either of the other two. This goal, therefore, has the best claim to be regarded as an elementary component of game playing. The others, since they are compounded components, can be defined only after the disclosure of additional elements.

The kind of goal at issue, then, is the kind illustrated by crossing a finish line first (but not necessarily fairly), having x number of tricks piled up before you on a bridge table (but not necessarily as a consequence of playing bridge), or getting a golf ball into a cup (but not necessarily by using a golf club). This kind of goal may be described generally as *a specific achievable state of affairs*. This description is, I believe, no more and no less than is required. By omitting to say *how* the state of affairs in question is to be brought about, it avoids confusion between this goal and the goal of winning. And because any achievable state of affairs whatever could, with sufficient ingenuity, be made the goal of a game, the description does not include too much. I suggest that this kind of goal be called the *prelusory* goal of a game, because it can be described before, or independently of, any game of which it may be, or come to be, a part. In contrast, winning can be described only in terms of the game in which it figures, and winning may accordingly be called the *lusory* goal of a game. Finally, the goal of participating in the game is not, strictly speaking, a part of the game at all. It is simply one of the goals that people have, such as wealth, glory, or security. As such it may be called a lusory goal, but a lusory goal of life rather than of games.

Means Just as we saw that reference to the goal of game playing admitted of three different (but proper and consistent) interpretations, so we shall find that the means in games can be of more than one kind – two, in fact, depending upon whether we wish to refer to means for winning the game or for achieving the prelusory goal. Thus, an extremely effective way to achieve the prelusory goal in a boxing match – viz., the state of affairs consisting in your opponent being 'down' for the count of ten – is to shoot him through the head, but this is obviously not a means for winning the match. In games, of course, we are interested only in means which are permitted for winning, and we are now in a position to define that class of means, which we may call *lusory* means. Lusory means are means which are permitted (are legal or legitimate) in the attempt to achieve prelusory goals.

It should be noticed that we have been able to distinguish lusory from, if you will, illusory means only by assuming without analysis one of the elements necessary in making the distinction. We have defined lusory means as means which are *permitted* without examining the nature of that permission. This omission will be repaired directly by taking up the question of rules.

Rules As with goals and means, two kinds of rule figure in games, one kind associated with prelusory goals, the other with lusory goals. The rules of a game are, in effect, proscriptions of certain means useful in achieving prelusory goals. Thus it is useful but proscribed to trip a competitor in a foot race. This kind of rule may be called constitutive of the game, since such rules together with specification of the prelusory goal set out all the conditions which must be met in playing the game (though not, of course, in playing the game skilfully). Let us call such rules *constitutive* rules. The other kind of rule operates, so to speak, *within* the area circumscribed by constitutive rules, and this kind of rule may be called a rule of skill. Examples are the familiar injunctions to keep your eye on the ball, to refrain from trumping your partner's ace, and the like. To break a rule of skill is usually to fail, at least to that extent, to play the game well, but to break a constitutive rule is to fail (at least in that respect) to play the game at all. (There is a third kind of rule in some games which appears to be unlike either of these. It is the kind of rule whose violation results in a fixed penalty, so that violating the rule is neither to fail to play the game nor [necessarily] to fail to play the game well, since it is sometimes tactically correct to incur such a penalty [e.g., in hockey] for the sake of the advantage gained. But these rules and the lusory consequences of their violation are established by the constitutive rules and are simply extensions of them.)

Having made the distinction between constitutive rules and rules of skill, I propose to ignore the latter, since my purpose is to define not well-played games but games. It is, then, what I have called constitutive rules which determine the kind and range of means which will be permitted in seeking to achieve the prelusory goal.

What is the nature of the restrictions which constitutive rules impose on the means for reaching a prelusory goal? I invite you, Skepticus, to think of any game at random. Now identify its prelusory goal: breasting a tape, felling an opponent, or whatever. I think you will agree that the simplest, easiest, and most direct approach to achieving such a goal is always ruled out in favour of a more complex, more difficult, and more indirect approach. Thus, it is not uncommon for players of a new and difficult game to agree among themselves to 'ease up' on the rules, that is, to allow themselves a greater degree of latitude than the official rules permit. This means removing some of the obstacles or, in terms of means, permitting certain means which the rules do not really permit. On the other hand, players may find some game too easy and may choose to tighten up the rules, that is, to heighten the difficulties they are required to overcome.

We may therefore define constitutive rules as rules which prohibit use of the most efficient means for reaching a prelusory goal.

Lusory attitude The attitude of the game player must be an element in game playing because there has to be an explanation of that curious state of affairs wherein one adopts rules which require one to employ worse rather than better means for reaching an end. Normally the acceptance of prohibitory rules is justified on the grounds that the means ruled out, although they are more efficient than the permitted means, have further undesirable consequences from the viewpoint of the agent involved. Thus, although nuclear weapons are more efficient than conventional weapons in winning battles, the view still happily persists among nations that the additional consequences of nuclear assault are sufficient to rule it out. This kind of thing, of course, happens all the time, from the realm of international strategy to the common events of everyday life; thus one decisive way to remove a toothache is to cut your head off, but most people find good reason to rule out such highly efficient means. But in games although more efficient means are – and must be – ruled out, the reason for doing so is quite different from the reasons for avoiding nuclear weaponry and self-decapitation. Foot racers do not refrain from cutting across the infield because the infield holds dangers for them, as would be the case if, for example, infields were frequently sown with land mines. Cutting across the infield in shunned solely because there is a rule against it. But in ordinary life this is usually – and rightly – regarded as the worst possible kind of justification one could give for avoiding a course of action. The justification for prohibiting a course of action that there is simply a rule against it may be called the *bureaucratic* justification; that is, no justification at all.

But aside from bureaucratic practice, in anything but a game the gratuitous introduction of unnecessary obstacles to the achievement of an end is regarded as a decidedly irrational thing to do, whereas in games it appears to be an absolutely essential thing to do. This fact about games has led some observers to conclude that there is something inherently absurd about games, or that games must involve a fundamental paradox.* This kind of view seems to me to be mistaken. The mistake consists in applying the same standard to games that is applied to means-end activities which are not games. If playing a game is regarded as not essentially different from going to the office or writing a cheque, then there is certainly something absurd or paradoxical or, more plausibly, simply something stupid about game playing.

But games are, I believe, essentially different from the ordinary activities of life, as perhaps the following exchange between Smith and Jones will illustrate. Smith knows nothing of games, but he does know that he wants to travel from A to C, and he also knows that making the trip by way of B is the most efficient means for getting to his destination. He is then told authoritatively that he may *not* go by way of B. 'Why not?' he asks. 'Are there dragons at B?' 'No,' is the reply. 'B is perfectly safe in every respect. It is just that there is a rule against going to B if you are

---

* See Chapter Seven, 'Games and Paradox,' for an extended discussion of this point.

on your way to c.' 'Very well,' grumbles Smith, 'if you insist. But if I have to go from A to C very often I shall certainly try very hard to get that rule revoked.' True to his word, Smith approaches Jones, who is also setting out for C from A. He asks Jones to sign a petition requesting the revocation of the rule which forbids travellers from A to C to go through B. Jones replies that he is very much opposed to revoking the rule, which very much puzzles Smith.

SMITH: But if you want to get to C, why on earth do you support a rule which prevents your taking the fastest and most convenient route?

JONES: Ah, but you see I have no particular interest in being at C. *That* is not my goal, except in a subordinate way. My overriding goal is more complex. It is 'to get from A to C without going through B.' And I can't very well achieve that goal if I go through B, can I?

S: But why do you want to do that?

J: I want to do it before Robinson does, you see?

S: No, I don't. That explains nothing. Why should Robinson, whoever he may be, want to do it? I presume you will tell me that he, like you, has only a subordinate interest in being at C at all.

J: That is so.

S: Well, if neither of you really wants to be at C, then what possible difference can it make which of you gets there first? And why, for God's sake, should you avoid B?

J: Let me ask you a question. Why do you want to get to C?

S: Because there is a good concert at C, and I want to hear it.

J: Why?

S: Because I like concerts, of course. Isn't that a good reason?

J: It's one of the best there is. And I like, among other things, trying to get from A to C without going through B before Robinson does.

S: Well, *I* don't. So why should they tell me I can't go through B?

J: Oh, I see. They must have thought you were in the race.

S: The what?

I believe that we are now in a position to define *lusory attitude*: the acceptance of constitutive rules just so the activity made possible by such acceptance can occur.

The definition

Let me conclude by restating the definition together with an indication of where the elements that we have now defined fit into the statement.

To play a game is to attempt to achieve a specific state of affairs [prelusory goal], using only means permitted by rules [lusory means], where the rules prohibit use of more efficient in favour of less efficient means [constitutive rules], and where the rules are accepted just because they make possible such activity [lusory attitude]. I also offer the following simpler and, so to speak, more portable version of the above: playing a game is the voluntary attempt to overcome unnecessary obstacles.

'Thank you, Grasshopper,' I said when he had finished speaking. 'Your treatment has completely cured my vertigo, and I believe I have a sufficiently clear understanding of your definition to raise a number of objections against it.'

'Splendid. I knew I could rely upon you.'

'My objections will consist in the presentation of counter-examples which reveal the definition to be inadequate in either of the two respects in which definitions can be inadequate; that is, they will show either that the definition is too broad or that it is too narrow.'

'By the definition's being too broad I take it you mean that it erroneously includes things which are *not* games, and by its being too narrow you mean that it erroneously excludes things which *are* games.'

'That is correct,' I answered.

'And which kind of error will you expose first, Skepticus, an error of inclusion or an error of exclusion?'

'An error of exclusion, Grasshopper. I shall argue that your account of the prelusory goal has produced too narrow a definition.'

For if no one had ever used his feet before the invention of racing, then foot racing would require the invention of

newfeld '78

running...'

4 Triflers, cheats, and spoilsports

In which the Grasshopper wards off an attack by Skepticus on his definition by distinguishing between a game and the institution of a game

Since many goals exist only because they are goals in *games* [I continued], it does not seem possible in these cases to identify a prelusory goal, that is, a goal which – in your words, Grasshopper – can be achieved 'independently of the game in which it figures or may come to figure.' How can checkmate, for example, be achieved aside from a game of chess? You cannot say, in an effort to dissociate checkmate from chess, that checkmate consists in objects of a certain physical description arranged in a certain pattern (or range of patterns), for it is the rules which govern movement of the pieces that permit an arrangement of the pieces to count as a case of checkmate, and not merely the geometrical pattern such an arrangement has. And the object of the game – to immobilize an opponent's king – is also a rule-governed state of affairs, since a king in chess is nothing more than a marker which is placed on the squares in accordance with rules. You expressly recognized this point earlier when you described the prelusory goal of chess in the following words: 'To place your pieces on the board in such an arrangement that the king is, *in terms of the rules of chess*, immobilized.' In chess, therefore, it seems impossible to identify a prelusory goal, that is, an achievable state of affairs which becomes the goal of a *game* only with the introduction of means-limiting rules. For the alleged prelusory goal of chess is already saturated with rules and is therefore not a prelusory goal as defined. I will grant you that more primitive games, such as foot racing, do have this kind of goal, for crossing a line drawn upon the ground can be accomplished independently of the rules of foot racing. But chess and a host of other games as well do not appear to have such goals. In this respect, therefore, the definition is too narrow.

Your objection is a good one, Skepticus [replied the Grasshopper], and you have expressed it with commendable force. I believe that it arises, however, out of a confusion of two quite different ways in which rules figure in games. Although it is necessary to refer to the rules of chess in describing checkmate, and also necessary, when playing chess, to obey the rules in seeking to achieve that state of affairs, the involvement of chess rules in the two cases should not obscure the fact that the uses to which the rules are put are quite different. In one case they are used to *describe* a state of affairs, in the other case to *prescribe* a procedure. And it is clear that one can avail oneself of their descriptive use without at the same time having to commit oneself to their prescriptive use. For one can bring about a state of affairs correctly describable as checkmate in complete disregard of the rules as procedural prescriptions. One simply sets out the pieces in such a way that Black has White in checkmate. Such 'descriptive' checkmate does not, of course, signal anyone's victory at chess, since it was not the result of anyone's playing chess. But for that very reason it confirms the contention at issue, namely, that there is a goal in chess which can be achieved independently of the game in which it occurs or may come to occur.

Still, while it is true that (descriptive) checkmate can be achieved without playing a *game* of chess, it is nevertheless the case that the achievement of such checkmate is in some sense dependent upon *chess*. What is this sense of chess? There is, I suggest, an *institution* of chess which can be distinguished from any individual game of chess. Because of this institution it is possible, for example, to take a knight out of a box of chessmen and describe its capabilities, even though the knight is not then functioning as a knight, that is, as a piece in a game of chess. And it is also possible, as has been noted, to set out the chess pieces in a checkmate arrangement without having to play a game of chess in order to achieve that state of affairs. Accordingly, although it is not possible to achieve the prelusory goal of chess (or at least to recognize that you have done so) aside from the institution of chess, it is possible to achieve it aside from a game of chess.

In order to lend further support to this conclusion, let us consider three familiar types of behaviour associated with the playing of games – the behaviour of triflers, cheats, and spoilsports. For it will be seen that the identification of these types presupposes the distinction between a game and its institution and the identification of a prelusory goal which that distinction permits.

A trifler at chess is a quasi-player of the game who conforms to the rules of the game but whose moves, though all legal, are not directed to achieving checkmate. Such a trifler may have some other purpose in mind. He may, for example, simply be trying to get six of his pieces to the other side of the board before he is checkmated, in which case he could be said to be trifling with chess by playing another game at the expense of chess. Or he may be interested simply in seeing what patterns he can create. Or he may just be moving his pieces at random. Now although it is possible for someone to do all of these things without violating the rules of chess, I think it is fair to say that such a person is not playing chess, although it is clear that he is operating within the institution of chess, for all he is doing is making chess moves. But to acknowledge the distinction between the game of chess and its institution is also to acknowledge the existence in chess of a prelusory goal, for it is the trifler's refusal to seek that goal which alone entitles us to say that although he is engaged in something chess-like, playing chess is not what he is engaged in.

Perhaps we can say of the trifler that he is not playing chess because of a deficiency of zeal in seeking to achieve the prelusory goal of chess. If so, then perhaps we can say of the cheat that he is not playing chess because of an *excess* of zeal in seeking to achieve the prelusory goal. For although, unlike the trifler, he certainly wants to achieve a condition which is, descriptively, a condition of checkmate, his desire to achieve that condition is so great that he violates the rules of chess in his efforts to do so. But he, too, is operating within the institution of chess, for he violates the rules in their prescriptive application only because of his expectation that they will be observed in their descriptive application. Thus, although he is not really playing the game, he has not abandoned the game's institution. On the contrary, his continuing to operate in terms of the institution is a necessary condition for his exploitation of the game and of his opponent. Liars, as Kant has pointed out, would soon go out of business if everyone were a liar, that is, if there were not a well-established institution of truth-telling. For if no one had more reason to believe than to disbelieve anything that anybody else ever said, then lying would not deceive and would so be pointless. In terms of their dependence upon institutions, cheaters at games are precisely like liars in everyday life. For suppose a cheat at chess has, without detection, feloniously achieved a pseudo-checkmate only to find that his opponent will not acknowledge that the checkmate arrangement of pieces counts as a victory.

'Checkmate,' says the cheat.

'Nonsense,' his opponent rejoins. 'Checkmate is the condition when you have immobilized my king. But you have *not* immobilized my king. Behold; I am moving it about in the air.'

'That isn't a move in *chess*, you idiot!' cries the enraged cheat.

'What rubbish. A move is a move.'

'Don't be absurd. How could I possibly counter such a "move"?'

'Why don't you try to grab me by the wrist?'

'How can you be so stupid? Do you want to play chess or do you want to arm wrestle?'

'Arm wrestle, now that you mention it. Chess bores me to death.'

'Damn you!' sobs the cheat. 'You're nothing but a spoilsport!'

'Bang in the gold,' replies the spoilsport.

In summary it may be said that triflers recognize rules but not goals, cheats recognize goals but not rules, players recognize both rules and goals, and spoilsports recognize neither rules nor goals; and that while players acknowledge the claims of both the game and its institution, triflers and cheats acknowledge only institutional claims, and spoilsports acknowledge neither.

The difference between chess and foot racing, therefore, is not that foot racing has an identifiable prelusory goal and chess does not; it is that foot racing does not – at least obviously – happen to require the kind of institution that is required by chess. In foot racing the 'moves' consist in kinds of running (pacing, sprinting, passing, etc.), and these already exist aside from foot racing in a way that the moves of bishops and rooks do not exist aside from chess. And the prelusory goal in foot racing – crossing a line ahead of other runners – does not require reference to the institution of foot racing in order to be intelligible. But even this difference between foot racing and chess is less sharp than at first appears to be the case. For if no one had ever used his feet before the invention of foot racing, then foot racing would require the invention of running, and so pacing, sprinting, and passing would be as much instituted moves as are the moves in chess. But this condition would not preclude identification of a prelusory goal, because the latter could be achieved – as in chess – by violating or ignoring the procedural rules which governed foot racing. And it is the latter fact which establishes the game-independence of a prelusory goal, not the fact that such a goal can exist outside the institution which includes the game.

'Well, Grasshopper,' I said, 'I can think of no immediate rejoinder to your reply. The distinction between a game and its institution seems to be undeniable, and therefore the universality of the prelusory goal as well. Let me, then, advance my second objection, which has to do with your characterization of constitutive rules. I shall argue that that characterization permits the classification of certain things as games which are manifestly not games.'

'Please proceed,' said the Grasshopper.

newfeld '78

'A game is when, although you can avoid doing something disagreeable without any loss or inconvenience you go ahead and do it any way.'

Taking the
long way home

In which the Grasshopper defends
his definition of games by intro-
ducing a definition of efficiency

Smith is at Jones's house [I began] and is about to depart on foot for his
own home. There are two routes he can take, a shorter or a longer one.
The shorter route is also the more scenic, since it takes him along the
cliffs and provides a spectacular view of Georgian Bay. The longer route
is across flat fields of stubble.

'I think I'll take the longer route tonight,' Smith announces to Jones.

'Why?' Jones responds.

'I've decided to make a game of getting home, you see.'

'Jolly good,' says Jones. 'What's the game?'

'I've just told you,' says Smith.

'Really? I must have missed it. Tell me again.'

'I said I was going to take the longer way home.'

'My dear fellow, that's not a game. It's a nuisance and a bore. The
view is monotonous and the path is overgrown with weeds and it will
take you longer.'

'Precisely. That's what makes it a game.'

'Afraid I don't follow you.'

'Let me try to explain. A game is when, although you can avoid doing
something disagreeable without suffering any loss or inconvenience,
you go ahead and do it anyway.'

'Who on earth told you that was a game?'

'A professor I know down at the university. He knows all about
games.'

'Obviously.'

'Actually, he didn't put it quite like that.'

'Aha.'

Smith then recites your definition to Jones, Grasshopper.

'It sounds a bit less idiotic put like that, I admit,' says Jones. 'Now suppose we take a look at this enterprise you want to call a game part by part and see if it fits the definition. First, are you trying to achieve some state of affairs?'

'Yes, the state of affairs which consists in my being home.'

'Right. Now, are there means at your disposal such that you can rule out some in favour of others?'

'Yes. I can go either the long way or the short way.'

'Check. Now, have you adopted a rule which prohibits more efficient in favour of less efficient means?'

'Clearly. I have ruled out going the shorter and therefore more efficient way.'

'And are you doing the latter just so that you can be getting home by taking the longer way around, and not for some ulterior purpose?'

'I am.'

'You don't have an amorous rendezvous in the meadow, do you, old chap? If you do then I wouldn't doubt that you were up to some game.'

'Unfortunately not. The facts are just as I have stated them.'

'Then one of two things must be the case.'

'And they are?'

'Either your decision to take the long way home is a game, or your professor is wrong in believing that voluntarily choosing less efficient over more efficient means is sufficient, with the other things he lists, to make what you are proposing to do a game. And since you will never get me to believe that your taking the long way home is a game, I can only conclude that your professor must be mistaken in his definition.'

'Yes, but suppose I actually do take the long way home for no purpose other than to be doing it. Surely I would be doing *something*, and not just nothing at all. If I am not playing a game, what am I doing?'

'As far as I can see, you would be doing something which you *believed* to be playing a game. What's the problem? Last night I believed that I was successfully bluffing Robinson into concluding that I had aces back to back, only I wasn't. I was making a mistake.'

If Smith's decision to take the long way home were a case of selecting inefficient over efficient means [replied Grasshopper], then my definition of game playing would be shown to be too broad, for I am entirely willing to admit that the activity Smith described to Jones is not a game.

But I shall argue that Smith's taking the long way home was *not* less efficient than his taking the short way home would have been. That task will require me to advance what I think is a fairly non-controversial definition of efficiency.

I define efficiency as the least expenditure of a limited resource necessary to achieve a given goal. I specify *limited* resource because if some resource is unlimited there is no reason to say that using more of it is less efficient than using less of it would be, *ceteris paribus*, regardless of the purpose or purposes for which it is used. My contention is that games exhibit inefficiency, so understood, but that Smith's taking the long way home does not.

Two examples should suffice to illustrate the fact that games exhibit the kind of limited resource necessary to characterize a use of means as being more or less efficient. Contestants in a foot race, for example, run fast either because they are competing against a record which limits the amount of time at their disposal (they do not have five minutes in which to run a four-minute mile) or against another runner whose pace limits the amount of time at their disposal. Their goal, that is, requires that they use as little time as possible. Since that is so, it can be said that running is a less efficient means for completing the course than, for example, riding a bicycle or driving a Ferrari. Resources other than time, of course, also figure in games. I earlier observed that the disposition of the pieces at a certain point in a game of chess might be such that a player's moving a knight along a diagonal would be a more efficient way to achieve a checkmate arrangement of the pieces than would be his moving the knight in the prescribed way. What is the limited resource which permits me to refer to reduced efficiency in that example? It is, I suggest, the number of moves a player has at his disposal. For if it will take Black fifty moves to checkmate White, and if it will take White forty-nine moves to checkmate Black, then Black will lose. Moving the knight along a diagonal, therefore, is a more efficient way of achieving the prelusory goal of chess just when such a 'move' has the effect of reducing the number of moves the player will need in order to achieve that goal.

Smith's taking the long way home is not a case of using less rather than more efficient means unless time, for example, is limited for Smith. It will not do to say that time is limited for Smith on the ground that time is *inherently* a limited resource. It is true that time is, alas, for all of us a finite quantity. But time, or any other potential resource, should not be called an actual resource unless it is viewed in relation to some goal, and it should not be called a limited resource unless it sets a limit, or limits, to the kind and number of my goals. Thus, although time is a finite quantity for everyone, it is not a limited resource for everyone. For a bored person time is a burden; for a person on the rack it is an agony. And when time *is* a resource for someone it is not always a limited resource. For a person with very few goals there is always enough time to accomplish all of them. And a classical Stoic, who on principle tailors his desires to fit his resources, must always in principle have just the right amount of time.

The trouble with the Smith counter-example, therefore, is that it is not, as described by Smith, a case which exhibits a selection of inefficient means, since there is no indication that any of the means that Smith proposes to use in getting home require drawing upon a limited resource. Smith's shoe leather *could* be a limited resource for him, or, more plausibly, his time. But it is perfectly possible that neither of these resources was limited for Smith. He may, whatever wear his shoes suffer, customarily buy himself a new pair every month. As for time, perhaps it lay heavy on his hands when he decided to set out for home. If so, then it would be much more to the point to say that Smith was operating efficiently in killing time than that he was operating inefficiently in getting home.

The counter-example may therefore be dismissed. But instead of doing so, let us recast the example in such a way that it will exhibit the kind of inefficiency that Smith's proposed 'game' lacked, for I predict that when it is properly revised the result will be accepted – by Jones and by you, Skepticus – as a game.

We must, then, stipulate that some resource relevant to his getting home is limited for Smith. Let us say that his time is limited. And to say that his time is limited is evidently to say that he has another goal which also requires time for its completion and that there is competition between these two goals for the time available. Let us say that Smith wants to get home before dark, that the sun has begun to set, and that the distance to his house is such that taking the longer way risks, to some extent, the outcome. Under these circumstances, it seems clear, taking the longer way is less efficient than taking the shorter way. And if Smith has no purpose in taking the longer route aside from his wish to engage in the activity such an obstacle makes possible, I submit that he is playing a game; specifically, he is having a race with the sun.

This example also provides me the opportunity, Skepticus, to reinforce a point I made earlier. In the original presentation of the definition

I pointed out that in his invention of games the gamewright must avoid the two extremes of excessive laxity and excessive tightness in the rules he is laying down, or else run the risk of aborting his invention. Thus Smith's 'game' was an abortive attempt because it did not contain any proscription of means, and the reason why it did not contain such a proscription was that time was not a limited resource. By extending the colloquy between Smith and Jones just a bit further we can also illustrate the opposite kind of failure. For this purpose, let us suppose that the two friends have formulated for themselves the same principle of efficiency that I have advanced, and have applied it to Smith's desire to make a game of getting home.

JONES: Why don't you take the long way home and try to get there *before sundown? That* would be a game, old chap.
SMITH: Impossible.
J: Why, for heaven's sake?
S: Because we've used up all the available game time in figuring out how to play the game. The sun has set.

'Very well, Grasshopper,' I said, 'I am satisfied that my objection to your account of constitutive rules on the ground that it makes the definition too broad has been satisfactorily answered. Let me, then, launch an attack once more from the other side, and argue that that account renders your definition too narrow – that is, that there are activities which must be acknowledged to be games which do not contain any limitation whatever upon the means which are permitted to achieve a 'prelusory' goal. I put prelusory in quotation marks because if the objection is sound the very distinction between lusory and prelusory goals must be abandoned.'

'This sounds a very serious objection indeed,' replied the Grasshopper.

'It is,' I assured him.

6 Ivan and Abdul

In which the Grasshopper defends his definition by entertaining and then rejecting the possibility of there being a game with no rules

Ivan and Abdul [I began] had been officers of general rank before each was retired and 'elevated' to the post of ambassador in the backwater capital of Rien-à-faire. Both had established brilliant military careers in the service of homelands which had frequently been at war with one another, and Ivan and Abdul had in fact been opposite numbers in many engagements. So the two warriors were overjoyed at the opportunity their appointments afforded them for going over all of their old campaigns together. But after a few months, when they had reviewed all the victories and defeats from every possible angle and refought all the old battles under every conceivable modification of logistics and tactics, they grew weary of their reminiscences and sought other diversions.

Sport seemed an obvious pastime for a couple of shelved warriors to take up, since sport seemed to them to be a kind of substitute, or polite, kind of warfare. It soon became evident to them, however, that sports were like warfare in only the most superficial respects. Specifically, they found that sports were hedged round with the most outrageously arbitrary restrictions. In golf, for example, you were expected to use a golf club to get your ball out of a sand trap even when your opponent could not see what you were doing. And in tennis, you were expected to call a ball foul or fair honestly even when your opponent was not in a position to check your call. Chess was no better, since surreptitiously to alter the location of pieces on the board – obviously an effective tactic – was ruled out.

But since they could find nothing better to do to occupy their time, they continued to play these games, although – as the diplomatic colony to its delight soon became aware – with a difference. Whenever the rules could be broken without detection or retribution, they were broken. Although this approach was ultimately doomed to failure, it worked very well for a time, and a number of breathtaking refinements were added to most of the conventional games. Thus to golf was added, among other things, the use of self-propelling radar-controlled golf balls, and to chess the use of hallucinogenic drugs as an offensive weapon. On the tennis courts Abdul achieved a much admired coup by hiring two men to raise and lower the net at appropriate times, until this was countered by Ivan's introduction of the net-piercing tennis ball. Things reached their fated conclusion in a climactic chess match.

In preparing themselves for the contest both contestants had countered the possible use of drugs by taking suitable antidotes, and each was determined to keep a very keen eye upon the other throughout the match. The first game proceeded normally for six moves. Then Ivan made the move which was the beginning of the end. Utterly ignoring the rules governing movement of the pieces, he illegally moved his queen to a square which put Abdul in check. The fascinated audience waited breathlessly for Abdul's response to this outrage. It was not slow in coming. He simply removed Ivan's queen from the board and put it in his pocket. Ivan in turn was quick to respond. In a trice he had nimbly rearranged the pieces on the board so that Abdul's king was in checkmate, crying, 'I've won!'

'Wrong, my friend,' screamed Abdul, and gathering up all of the pieces except his king, he flung them to the floor.

'Abdul, you can't do that,' said Ivan in outraged tones. 'I won the game the moment you were in checkmate.'

'So you say,' responded Abdul, 'but you were obviously mistaken, for there stands my king, quite free to move.'

Ivan had not, of course, expected such a transparent tactic to succeed with the wily Abdul. It had merely been a diversionary move so that he could, while his opponent was momentarily distracted, secure Abdul's king to the board with the quick-drying glue he had all along held ready in his hand beneath the table. Then, of course, before you could say 'scimitar,' Abdul snatched a bottle of solvent from his tunic and freed his king. Ivan's hand immediately shot out towards the king, but Abdul grabbed his wrist in time to forestall the assault. For a full minute they were locked in a frozen tableau of force and counterforce (evoking spirited applause from the audience), before they broke apart, leapt from their chairs, and began to circle each other warily. Then they joined battled in what was to become a truly mythic contest, for

> They fought all that night
> Neath the pale yellow light,
> And the din it was heard from afar.
> Huge multitudes came,
> So great was the fame
> Of Abdul and Ivan Skavar.

The legend then incorrectly goes on to recount the game as ending in a tie with the mutual destruction of Abdul and Ivan, followed by some sentimental reference to a tomb rising up where the blue Danube flows and to a Muscovite maiden her lone vigil keeping 'neath the light of the pale polar star, but that is all the most obvious kind of bardic invention and ornamentation. The game did not end in a tie, but in a stalemate, when both fell to the floor in utter exhaustion, unable to move, and when it was discovered that one of the spectators had made off with the board and the pieces.

In fact, the two friends met the following afternoon at their favourite cafe. Said Ivan, 'My friend, that was the best chess game I've ever played.'

'Oh, unquestionably,' replied Abdul.

They drank their aperitifs in companionable silence. Then Ivan spoke again.

'Still, there is something that bothers me.'

'Indeed,' said Abdul, 'Perhaps, you know, the same thing is bothering me.'

'I shouldn't be surprised. If you are thinking what I am thinking you will have realized that it will be impossible for us ever to play chess again.'

'Just so. The instant of setting out the pieces for a game would be the signal for us to start a battle whose weapons had nothing whatever to do with chess, since the only moves either of us will accept are moves that really coerce, either by force or by deceit. For, since we will not abide by the rules of the game, the winner can be only he who has gained final mastery of the situation. And of course, it's not only that we can no longer play chess. For the same reason, we can no longer play any game, for games require that we impose artificial restraints upon ourselves in seeking victory, and we refuse to do that.'

'Exactly,' said Ivan. 'When I had my brigade and the general staff used to issue their namby-pamby orders in the name of military honour, I swore that if ever I was chief of staff I would root out all that kind of thing. Rules of war indeed!'

'What about gas? You wouldn't have used gas, would you?'

'Of course not. But not because I wanted to "play the game." Gas is too risky for the user, as you well know, and besides that only an idiot would intentionally invite that kind of retaliation. The same with nuclear bombs, of course. Refraining from leading with your chin isn't chivalry, it's basic strategy.'

'Still, artificial restraints do have their uses. Oh, not in war, *mon vieux*, I agree with you there. But an awful lot of people do seem to play chess and golf, you know, without getting into a brawl.'

'Civilians, old boy, *civilians.*'

'What about all the officers who play golf at the country club?'

'Jumped up civilians. Good candidates for the general staff.'

'Still, Ivan, look at all we're missing. I sometimes wish I could play by the rules.'

'Wishes don't cost anything, Abdul. The question is, *can* you play by the rules?'

'I suppose not.'

'Of course not. We are what we are.'

'Then it looks as though we'll have to go back to reliving our past glories for the rest of our days. Maybe it's time just to pack it in, Ivan, as a noble Roman would have done.'

'I don't think it has quite to come to that, my friend.'

'You have an idea, Ivan, I can tell.'

'A germ, Abdul, a germ. I'm going to sleep on it, however. Tomorrow at the same time?'

'Very well. Till tomorrow.'

Next day Abdul found his friend already seated at their table at the cafe smiling broadly at the tumbler of vodka before him.

'Tell me your idea at once, Ivan,' said Abdul, seating himself at the table.

'At once, my friend, at once. I have thought about it all night and most of the day, and I am satisfied that the logic is absolutely compelling. There is one, and only one, game left for us to play.'

'What game, Ivan? What logic?'

'A fight to the finish, my friend.'

'What! Ivan, you must be mad!'

'On the contrary. It is demonstrably plain that any other alternative would be imbecilic. We have seen that for you and me no game can be won by either of us unless he has complete mastery over the other. We cannot add, as civilians do, "complete mastery *in terms of the game*" because that means in terms of the *rules* of the game, and we do not acknowledge such rules. Thus, the other night, when I in defiance of the rules summarily arranged the pieces so that your king was in the position of being checkmated, the civilians would say that I had not really won the game because I had not achieved that state of affairs by following the rules, is that not so?'

'Yes, certainly.'

'And we, too, found that I had not won the game, but for a quite different reason, *n'est-ce pas*?'

'That's right. You had not won the game because you were unable to hold your position.'

'Yes. So we may say that when civilians win games they are always looking to the past, for all they care about is how they got there, but for us, once we have achieved a success what matters is not how we did it, but whether we can sustain our position. We are always looking to the future.'

'That's quite well put, Ivan.'

'Yes. And that is why the only kind of game we can play, Abdul, is a fight to the finish.'

'I'm afraid I don't quite see *that*, my friend.'

'Well, we are agreed, are we not, that for you and me victory consists in mastery of one of us over the other, regardless of the game that is being played?'

'Yes, we are agreed on that.'

'Well, then, Abdul, let me ask you this. In any game we choose to play – or in *the* game, since there can be only one game for us – how long must one of us have mastery over the other for such mastery to count as winning?'

'Well, Ivan, why couldn't we just assign an arbitrary time limit? Five minutes, a day, a week, it doesn't really matter, does it?'

'Abdul, Abdul, you're not thinking. Your solution of the problem posed by the fact that you and I cannot play rule-governed games is to invent a rule. What kind of solution is that?'

'Yes, I see. That is, if my suggested time limit is in fact the same as a rule.'

'But isn't it perfectly clear that it is? I immobilize your king for, let us say, five minutes by gluing it to the table and holding you at bay with a revolver so that you cannot apply your solvent. At the end of five minutes I pocket my gun and declare myself the winner. Surely you're not going to tell me, my friend, that your response would be to congratulate me on a game well played?'

'No, Ivan, I am not. I would immediately draw my weapon and hold you off while I applied the solvent.'

'Of course you would, because for us a past victory is worthless unless it can be extended into future domination.'

'So the answer to the question how long one of us must dominate the other is that it must be for ever.'

'Just so. And since by "domination" we mean freedom from attack by the one dominated, it is clear what efficiency in achieving domination, if I may put it that way, demands, is it not?'

'It is. No one can be sure that he is safe from attack by an opponent unless the opponent no longer exists to attack him.'

'Therefore, my friend, since we know that we cannot play a game that has rules, it follows that if we are to play a game at all, it must be one without rules, and a fight to the finish is the only game without rules that there is. Q.E.D.'

'I agree. But are you serious in suggesting that we act on this conclusion?'

'I am entirely serious. What are the alternatives? You yourself, just yesterday, were entertaining the possibility of committing suicide. Is that preferable?'

'No, it isn't.' There is a thoughtful silence, at length broken by Abdul with a laugh.

'What is it?' said Ivan.

'I was just thinking. The French are supposed to be the most logical thinkers in the world, but I think only you Russians, Ivan, are crazy enough to act on the basis of a cogent chain of reasoning no matter where it leads.'

'Then you do not wish to play this ultimate game?'

'On the contrary, I am quite prepared to play it. It is just that, if I had been left to myself, I doubt that I am kinky enough to have actually made the final commitment.'

'Yes, well, that is why the world has never heard of Turkish Roulette.'

'No, nor Russian Delight either. But tell me, have you yourself ever played Russian Roulette?'

'Not lately. The general staff and the Foreign Office frown on general officers and ambassadors amusing themselves in that way. But as a subaltern I used to play it all the time.'

'And are alive to tell the tale! You must have been fantastically lucky.'

'Luck had nothing to do with it. I always palmed the bullet. But enough of this. I am keen to begin the game. Will you be ready to start at dawn tomorrow?'

'Quite ready.'

'Then, since each of us no doubt has some preparations to make, I will take my leave of you. Abdul, farewell.'

'Farewell, Ivan.'

If [replied the Grasshopper] Ivan's and Abdul's proposed fight to the finish is a game in which there are no rules that prohibit more efficient in favour of less efficient means, then my definition must be too narrow. The definition can be defended, therefore, only if the fight to the finish is not a game or else has, in fact, the kind of rules the definition requires. And since I am quite willing to accept that their fight to the finish *is* a game, evidently I must show that, unbeknownst to Ivan and Abdul, their game did indeed contain at least one rule of the required kind. And I believe that I can show just that, simply by asking you, Skepticus, to consider the following question: 'Why didn't Ivan destroy Abdul immediately upon committing himself to a fight to the finish with him?' He could easily have done so while they were talking things over in the cafe, but he did not. Instead, quite unaccountably, he proposed to Abdul that the game begin at dawn on the following day. Let us awaken Ivan just before dawn on the appointed day and put this question to him.

'Ivan, are you awake?'

'I am. Who is it? What do you want?'

'I am the Voice of Logic, and I have a question to put to you.'

'What time is it?'

'An hour before dawn.'

'Put your question, then, but please be brief.'

'The question is a short one. Why didn't you destroy Abdul just as soon as you had decided to have a fight to the finish with him?'

'Here is an equally short answer. Because I have no interest in destroying Abdul *per se*. I am interested in seeking to kill him only so that I can be battling him.'

'Let me test that allegation, if you don't mind.'

'Test away.'

'Very well. I tell you that Abdul is at this moment fast asleep in his bed. You can easily gain entrance to the embassy and kill him in his sleep, thus winning the battle with a minimum of risk by a stunning surprise attack.'

'As you can see, I am not leaping from my bed and speeding to the embassy.'

'Yes, I do see that, and it puzzles me very much.'

'I don't see why it should. If I kill Abdul before the game starts, then I can't very well fight him, can I? If I killed him now, our game could never begin.'

'You are saying that this game you are going to play has a starting time.'

'Of course.'

'In other words, there is a rule which forbids you to make a move in the game before a certain agreed upon time.'

'A rule, you say?'

'Yes,' responded the Voice of Logic inexorably, 'a rule.'

'Then,' said Ivan, frowning and sitting up in bed, 'our fight to the finish is not really a game without rules.'

'Not if you stick to your dawn starting time.'

'And I thought we had finally found a game without the artificiality of rules. How could we have missed this business of a starting time?'

'Perhaps it was because you were so busy eliminating an ending time. But it is perfectly clear, is it not, that a starting time is just as much an artifice as a finish time?'

'Yes, it is.'

'And now that you know this, you will of course at once sneak up on Abdul in his sleep and kill him, right?'

'Not at all.'

'Why not?'

'I have answered that question twice already. Damn it, I don't want to murder Abdul – I like him, for God's sake – I just want to play a game with him.'

'Yes, I understand that. And you also want to play a game without rules that artificially limit the means at your disposal for achieving victory. Isn't that correct?'

'Yes, it is.'

'Well, now that you see that a starting time is such an artificial limitation, why don't you play this game that you have at last correctly formulated, and go and kill Abdul?'

'Because with the elimination of a starting time I have also eliminated the possibility of there being a game at all.'

'Why is that?'

'Well, the game Abdul and I propose to play is a contest, is it not? But the game cannot be played unless the contestants actually contest. Killing Abdul in his sleep would be just like slaughtering an opposing football team before they reached the stadium and then claiming that you had won the match.'

'So accepting the limitation of a starting time is the same as ensuring that you will have an opponent; that is, someone who is prepared to attack you as you are prepared to attack him.'

'Precisely. If that were not inherently part of the idea of a competitive game, then I might just as well have killed Abdul two days ago, before the idea of a fight to the finish ever occurred to me, or, for that matter, I might just as well have killed some chance person and then claimed that I had triumphed in a fight to the finish. But there is no victory in killing some unsuspecting victim. Anybody can do that.'

'In other words, just killing Abdul does not count as winning the game, for that goal can also exist aside from the game, as in the case of murder.'

'Yes. Winning consists in killing Abdul only under conditions that mean he is also in a position to kill me, and where both of us know that it is kill or be killed. That is the whole meaning of an agreed upon starting time.'

'Would you agree with the following general account of what you have just said? You are attempting to achieve a certain state of affairs (the death of Abdul), using only means permitted by a rule (both of you must know at the same time that each is out to kill the other), where this rule prohibits more efficient in favour of less efficient means (it would be much more efficient to achieve the death of Abdul without issuing a challenge and receiving an acceptance), and where – a point not now in dispute – the sole reason for accepting the limiting rule is to make possible such activity.'

'Yes, that describes the situation perfectly.'

'Well, if you are prepared to play such a game, I don't see why you aren't prepared to play *any* game. If, that is, you are prepared to accept what might be called an unnecessary obstacle in order to be able to play this game with Abdul, why not accept other unnecessary obstacles and play chess or tennis or golf with Abdul instead, and give up this folly of a fight to the finish? Either that, or admit that there is no reason to wait for the starting signal and kill Abdul now.'

There is silence as Ivan turns this over in his mind. Then he leaps from his bed, flings on his clothes, and rushes wildly from the room.

'Where are you going?' cries the Voice of Logic.

'I must reach Abdul before dawn!' cries Ivan from the staircase.

'To call off the game or to kill him?' disjunctively queries the Voice of Logic.

But Ivan's shouted reply is too muffled to understand as he rushes pell-mell through the dark and deserted streets.

Nearly half way to Abdul's embassy Ivan sees a figure approaching at the opposite end of a short boulevard. It is Abdul. Has Abdul, too, been listening to the Voice of Logic? And is he hurrying to Ivan to call off the game, or to make a surprise attack? If Ivan can be sure that Abdul is making a surprise attack, then it is no surprise and the game can begin, for it has gained a starting time and the time is NOW. But how can Ivan be sure that it is NOW unless he knows what Abdul's purpose is? And Abdul may, of course, be in the same quandary. Ivan might shout, 'Let's call off the game!' But Abdul might very sensibly take this to be a ruse on Ivan's part for gaining an advantage. And Ivan, if Abdul called out the same proposal to him, would be foolhardy indeed to accept it out of hand as a genuine offer. Both stop in perplexed indecision.

And there they stand to this very day, in the form of two marble statues facing one another along the length of the Boulevard Impasse in the capital city of Rien-à-faire. At least that is the story the guides of Rien-à-faire tell to explain the sculptured confrontation along embassy row.

'I must say,' I said with a laugh when the Grasshopper had finished, 'you have fitted my illustrative tale with a startling denouement.'

'To be sure. But the question is, Skepticus, are you persuaded that all games must have the kind of constitutive rule which the definition requires?'

'Almost, Grasshopper, almost.'

'But not quite?'

'Well, even if the saga of Ivan and Abdul shows that *competitive* games must have at least one such rule, it does not show that non-competitive games are bound by the same requirement, and presently I shall raise an objection on that score. First, however, I am compelled to raise an objection which, if sound, would appear to undercut and render futile any attempt whatever to give a rational account of competitive games, thus making your present defence of constitutive rules irrelevant even if it is not unsound.'

'Good heavens, Skepticus, is there such an objection?'

'I'm afraid there is, Grasshopper. For it has been argued that competitive games are fundamentally paradoxical undertakings. And since to be paradoxical is, I take it, the same as to be inexplicable, it would seem to follow, if the argument is cogent, that a definition of competitive games is impossible.'

'Yes, that would surely follow. For when we discover that something we have been trying to understand is really a paradox, then reason compels us to abandon the quest, just as it would if we were seeking perpetual motion or a merciful banker. But I wonder, Skepticus, if you are thinking of Aurel Kolnai's address to the Aristotelian Society titled "Games and Aims," in which he argues for the thesis that you have suggested.'*

'That is so, Grasshopper. I have just finished reading it in the Society's published *Proceedings*. And it seems to me that Kolnai makes a rather plausible case for his position.'

'Yes, well, as it happens I too have read the piece and have, in fact, prepared a brief response. In fact, I have it with me, for I intend to send it off to the *Entomological Review*. If you would like to accompany me to the post office, perhaps I could read it to you on the way.'

'By all means, Grasshopper.'

---

* Aurel Kolnai 'Games and Aims' *Proceedings of the Aristotelian Society* 1966, 103–28

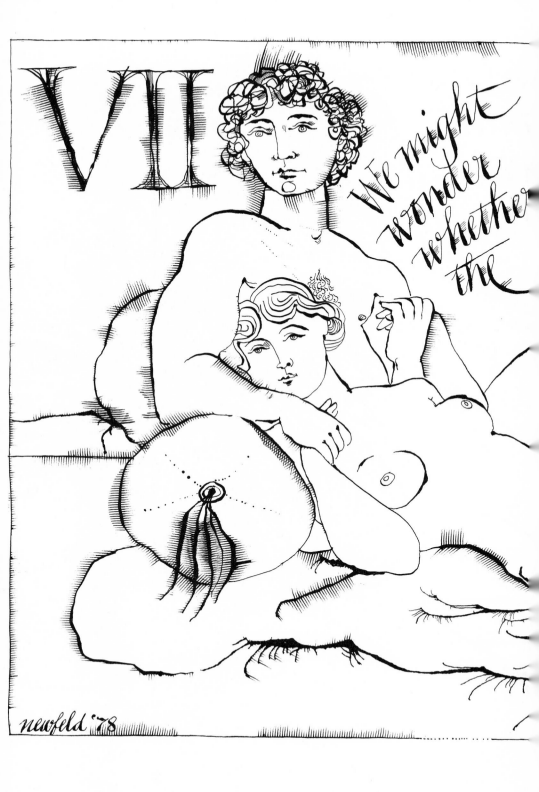

sexual act
must be
considered
a type of game.

7 Games and paradox

In which the Grasshopper examines and then rejects the possibility that competitive (zero-sum) games are fundamentally paradoxical and thus presumably indefinable

In his address to the Aristotelian Society [the Grasshopper read] Aurel Kolnai suggests that games exhibit what he calls a 'genuine paradoxy.' I do not believe that he has shown this to be the case, even on the most liberal interpretation of what it means to be a paradox. He has, however, called attention to an aspect of games which invites further investigation, and I should like to advance the following considerations not so much as a criticism of Kolnai as an attempt to take the investigation along a path which Kolnai has indicated, but which he has not himself, in my opinion, followed.

Kolnai's statement of the alleged paradox is as follows:

... the indissoluble double purposiveness of playing chess in absolute concord for the common pleasure of it and each player *in* chess aiming at *nothing but* defeating the other, destroying his power and foiling his purpose is what to me seems to exhibit in boldest outline the odd volitional posture I have ventured to call the paradoxy of Game (p. 112).*

If a genuine paradox involves an inescapable contradiction, then Kolnai has not shown that games exhibit a genuine paradox, for what

---

* It should be noted that Kolnai makes this claim only with respect to zero-sum games, which he describes in the following way: 'I call "unproductive" or "zero-sum" a type of game not implying any measurable or recordable achievement other than winning and losing or gains and losses between the partners, in terms of the game as such' (p. 109). My discussion of paradox, too, is confined to zero-sum games so described.

Kolnai advances as a contradiction can be escaped. I take it that Kolnai regards the posture in question to be volitionally odd because he sees any game player, *qua* game player, as possessing two incompatible aims. The two aims are 1/ an aim at concord between A and B (where A and B are competing players in the same game), and 2/ an aim which entails the negation of concord between A and B. Let us call the concord which seems to be at issue C. Then the 'contradiction' which seems suggested would consist in the players aiming at 'C and not-C.'

It seems perfectly clear, however, that the apparent force of the paradox depends upon an equivocation on C. The 'common pleasure' of the game, with respect to which the players are in 'absolute accord,' is, to be sure, a pleasure which arises from each player attempting to defeat the other. They are, therefore, in 'absolute accord' that there be competition between them. And of course they are *not* in accord with respect to the desired *outcome* of that competition, for if they were then the competition itself could not occur. But it is clear that their possession of these different aims does not entail a contradiction, for the 'concord' which both affirm is different from the 'concord' which both deny. And so the proposition that they are in concord about who will *compete* but that they are not in concord about who will *win* is so far from being paradoxical as to amount to a tautology; that is, its *denial* would produce a paradox. Kolnai's alleged paradox, therefore, is resolved by seeing that A and B are not, in fact, aiming at 'C and not-C' but at 'C_1 and not-C_2,' and, indeed, at 'C_1 *only if* not-C_2.' That is, their disagreement as to the desired outcome of the conflict is a necessary condition for there being any conflict at all.

Kolnai himself evidently sees this very point, although he seems to regard it – paradoxically, if I may say so – not as the resolution of the paradox but as the paradox itself. He states that 'to play chess and to mate one's partner ... are complementary and mutually conditioning pursuits' (p. 104). He continues:

the player's primary aim is to play chess rather than to win: but so as to attain that primary aim, he must by definition set up and pursue the sharply different aim of mating his opponent or at least frustrating his opponent's aim of mating him. That the two inseparable 'aims' are sharply different is obvious from the fact that the primary aim is necessarily common to both partners, whereas the implied aim of 'winning' is just as necessarily split into two antagonistic aims, one player's victory being identical with the other's defeat.

Kolnai evidently believes that the difference, indeed 'sharp' difference, between the aims of playing and winning is the basis of the alleged paradox. Yet it is just the fact that the co-operative and antagonistic aims of the players are directed to different ends rather than to the same end which renders such a volitional posture non-paradoxical.

It may be, however, that I have imposed too restrictive a meaning on Kolnai's use of the term 'paradox.' Perhaps he takes it to mean not inconsistency but only apparent inconsistency. But Kolnai insists on calling the paradox in question a genuine paradox, as though he thought it did indeed involve a real contradiction. Possibly he means by 'genuine' that the paradox genuinely seems to involve a contradiction. But, of course, seems to whom? Still, leaving aside the fact that some pseudo-contradictions are likely to seem genuine to some observers but not to others, I am prepared to offer the ease with which Kolnai's paradox is resolved as evidence against its even genuinely seeming to be a contradiction. A paradox which depends upon an easily discovered equivocation is not, one would think, even much of a seeming-contradiction.

By way of contrast, consider the following case. If players in games were found to be both co-operative and antagonistic with respect to the *same* end, this might well warrant our calling the joint possession of such aims paradoxical. Thus, if a player were to aim at both obeying the rules (in order to play the game) and breaking the rules (in order to achieve a quasi-victory, or perhaps the cash prize), we would recognize this as a genuine conflict between co-operation with and antagonism to the other player. But although this might be called a genuine paradox – the Paradox of the Schizophrenic Cheat, perhaps – one would not want to identify it as the odd volitional posture characteristic of games, which is not quite so odd as that. Indeed, the volitional posture normally characteristic of games is not sufficiently odd to qualify as a genuine paradox at all, since games do not require us to adopt conflicting intentions, but simply to intend conflict.

My quarrel with Kolnai, therefore, is not that the paradox which he believes to characterize games is resolvable, but that it is too readily resolvable. Nor do I object to anyone's calling something paradoxical even though what is involved is merely an appearance of contradiction. To point to such a 'paradox' is one way to express that wonder which Aristotle suggests is the beginning, but not the end, of philosophical inquiry. Kolnai's paradox, then, I find deficient in two respects. By evidently regarding the paradoxy of games as the end rather than the beginning of inquiry he seems content to leave us in a state of wonder about games,* and, since the paradox is readily resolved, the wonder in which we are left is not all that wonderful.

What we want, I should think, are not genuine paradoxes but fruitful paradoxes – oddities which lead us to discover that it is not the oddity but its denial which is genuinely odd, with the result that we learn something about whatever it was which at first seemd odd. ('There is nothing which would surprise a geometer so much,' said Aristotle, 'as if the diagonal [of a square] turned out to be commensurable.') A consideration of games does, I believe, disclose paradoxes of this kind.

One such paradox is suggested by the case of the schizophrenic cheat, which involved conflicting aims with respect to playing the game; viz., both fairly and unfairly. Alternatively, a person might harbour conflicting aims with respect to winning the game. One might aim both to defeat an opponent and also to co-operate with that opponent in his efforts to defeat oneself. This might be called the Paradox of the Reluctant Victor. Is such a set of aims a feature of games? Sometimes it is. Consider a game in which the players are very poorly matched; for example, a novice at chess against an experienced player. The novice is about to make a move which would enable his opponent to mate him in two additional moves. The experienced player points this out to the novice, the novice moves more effectively against his opponent, and the game continues. Now, even though the experienced player appears to be exhibiting contradictory intentions – he aims at defeating his opponent but intentionally puts obstacles in the way of doing this – we do not find his behaviour unintelligible or even irrational. That, of course, is the

---

* This statement may require additional support. In a later section of his paper, Kolnai contends that the relation between the aims of winning and playing is a very special kind of relation: 'I propose to call it the *paratelic* type of relationship, seeing that the internal or thematic aim – "winning" or "mating" ... – may be looked upon as a lateral implicate of the enveloping or primary aim of "playing, etc.," as a secondary but integral and somehow autonomous aim generated by the prior decision of "engaging in this game" ... ' It may be that Kolnai regards the paratelic relation as a resolution of the 'paradoxy of Game.' However, 1/ he does not tell us that he does, and 2/ if he does so regard it he would seem to be mistaken. If the aims of winning and playing are related as reciprocally necessary conditions, then it is hard to see how one of them could be 'prior' to and 'generate' the other, as though one could separate 'aiming to win' from 'aiming to compete' (which is what 'playing' means in a zero-sum game). For the latter may be translated 'aiming to defeat an opponent,' and that is identical with 'aiming to win.'

oddity, or, if you like, the paradox. Why do we not find this effort to defeat one's own purposes odd? The answer lies in the fact that this particular effort of the kind 'to defeat one's own purposes' is being made with respect to a game. We are thus led to ask what there is about games which renders such behaviour non-paradoxical, the implication being that aside from games such behaviour would be paradoxical, as in the following example. A general, A, aims to defeat the enemy, and has victory within his grasp provided the enemy does not obtain certain intelligence about the placement of troops. Now A, who really aims to defeat the enemy, intentionally provides the enemy with this intelligence. We try to find out why he did so and we find, let us say, that his sole reason was his wish not only to defeat the enemy but also to co-operate with the enemy's aim to bring about his own, i.e., A's, defeat. We might then conclude that we had discovered a person so fundamentally good-natured that he could not bear to disappoint anyone. And we might observe that such unbounded good nature is likely to produce intentions which indeed lead to odd volitional postures, and that these postures involve, in a manner quite different from the posture Kolnai attributes to game players, a basic paradoxy of attitude. We might call such a paradox the Paradox of Infinite Benignity.

In games, however, the peculiarity of giving strategic information to an opponent is not quite that peculiar. The superior chess player cautions his inferior opponent against a bad move not because he wants his opponent to win, but because he wants his own eventual victory to be more satisfying. He wants to win but he does not want to win, let us say, too soon. In the same way the general's behaviour would be rendered non-paradoxical if his aim in providing the enemy with valuable strategic information were not the enemy's victory, but the war's prolongation. Generals might, and perhaps often do, value both combat and victory as ends in themselves. In Kolnai's terminology, as this applies to games, both playing the game and winning the game are, in addition to being reciprocally necessary conditions, 'autotelic' aims. It is this feature of games which resolves the paradox.

But it might seem that games have escaped the Paradox of Infinite Benignity only to be caught up in a paradox of another kind. Thus if, given two aims, the achievement of one can hinder achievement of the other, then those aims must be in some sense opposed to one another, so that there may be said to be a paradoxy in the attitude of the person who holds both of them. This kind of thing can happen in games: 1/ the aim of winning, if it is accomplished too easily, thwarts achievement of the aim of playing the game, and 2/ seeking to achieve the end of playing the game (e.g., prolonging the game by helping an opponent in his efforts to defeat you) may thwart the aim of winning it. These paradoxes may be called, respectively, the Paradox of the Compulsive Winner and the Paradox of the Procrastinating Player.

Still, holding aims which may conflict is quite different from holding aims which necessarily do conflict. The aims of winning the game and (satisfactorily) playing the game do not necessarily, or even usually, come into conflict. Rather, as was suggested by the example of the novice and the experienced player, such a conflict is apparently charac-teristic of games which are in some sense defective, either 1/ because the players are poorly matched, or 2/ because the game itself is so con-structed as to make it likely that one of the players, though not superior to his opponent, will gain an unassailable advantage over the other; e.g., if the player who moved first were always or usually to gain such an advantage. We may say that conflict between the aims of winning and playing can occur where a game or the play in a game is inferior, and that the occurrence of such conflict is a sign that the game or the play is inferior. Correspondingly, a good game is one in which, for the winner, the aims of playing and winning are jointly realized, perhaps in terms of some kind of optimal balance. That is, a good game is just the kind of game which avoids the 'paradox.' And perhaps one could capture a basic feature of games in terms of 'paradox,' therefore, not by claiming that games exhibit a basic paradoxy, but that games are the kind of thing in which the possibility, indeed the danger, of such a paradox can occur. Thus, only to the extent that the occurrence of such conflict endangers the activity is the activity a game, or better, perhaps, game-like. Good games, it might be said, are just games which successfully avoid this paradox. Whether one wishes here to use the term 'paradox' is not of too great importance; 'conflict' will do, or 'conflict of intentions,' so that we may say that well-played games are just those which avoid what would otherwise be a genuinely odd volitional posture.

The relation between playing games and winning games seems to be exhibited more generally in a class of activities which may be called 'trying and achieving' activities of a special kind; namely, where the trying and the achieving are each sought as ends in themselves. Games, it should be noticed, are not the only activities of this kind. 'It is better to have loved and lost than never to have loved at all.' What might be called the standard (though admittedly not the only) sexual act is perhaps not only the intrinsically most interesting activity of this kind, but also the clearest case of this kind of activity. Trying to have an orgasm and having an orgasm are, I should imagine, rather widely regarded as each an end in itself, so that the achievement of one may thwart achievement of the other. Though orgasm is an end in itself, to achieve it instantaneously is to defeat the aim of building up to it. And, notoriously, to attend too single-mindedly to the build-up can preclude the orgasm. We might, accordingly, wonder whether the sexual act must be considered a type of game. To see that it need not be, we might begin by noticing that the essential effort in the sexual act we have described seems to be just the effort to balance the trying against the achieving. The effort in a game, however, is simply to win the game, and if the game is well constructed and the players are well matched, then the desired balance between trying and achieving will be realized, although such balance need not (and perhaps should not) be the thing to which the players' efforts are consciously and primarily directed. In playing well-constructed games well, that is to say, one is aiming to defeat an opponent, not a paradox.

But this will not really do as a difference between game playing and sexual activity. With respect to well-matched sexual partners, too, the balance between trying and achieving may (and perhaps should) be realized without conscious effort being directed towards its achievement. The case of the novice and the expert is as fitting here as it is in games.

Yet I want to maintain that playing games is different from sexual activity, and to that end I would like to propose a final 'paradox' about games: in games losing is achieving. Consider a sexual effort in which orgasm is not achieved. This is not like losing the game, because losing the game implies that someone else has won the game, whereas failing to complete the sexual act does not imply a winner. Or, if 'nature' (in the form of the physiological limitations imposed upon human beings) is regarded as an opponent who has 'won' by successfully frustrating the lovers' joint effort to gain a victory over it in the form of optimizing the balance between trying and achieving, then this would be to regard the sexual act as a game with 'nature.' But where the sexual act is not so regarded, failure to complete it is not like losing the game but like failing to complete the game: e.g., if a baseball game were to go into so many extra innings that both teams gave up the whole thing for ever as hopeless. The point is that one can complete a game by losing as well as by winning it. In losing a game, one has achieved something, even though one has not achieved victory. Has one achieved losing? To say that I have achieved the loss of the game seems the same as to say that I have succeeded by failing, and to say this of most types of 'trying and achieving' activities would be truly paradoxical (the Paradox of the Unbeatable Loser?). Why, then, is it not paradoxical, if it is not, to say this of a game? Because losing is only one way in which one can fail to win a game. One can also fail to win if the game is called off (for good) because of rain, or if it continues so long without a victor that the attempt to decide a victor is given up, or if one is disqualified because of cheating, or if one is struck dead before the end of the game. But failing to win the game by virtue of losing it implies an achievement, in the sense that the activity in question – playing the game – has been successfully, even though not victoriously, completed. In the case of the sexual act that we are considering, however, it does not make sense to say that one has successfully completed the activity but did not have the orgasm. And if it were to make sense to say that, then this would appear to indicate that the instance of sexual activity about which it made sense to say it was, in fact, a game. Thus one partner might say to the other, 'You won that time,' or both might say, with respect to nature, 'We lost that time, but it is better to have loved and lost than never to have lost at all.'

The ability to achieve a loss is not, *in games*, paradoxical. Nor is it odd, in the sense of being inexplicable. It is itself an explanation of a feature of games. This feature, to be sure, might be called odd. Still, not odd in itself, but only when compared with other activities, such as sexual activity, and then odd only in the sense of 'different from.' But to see this is to see that the feature is in another sense not odd at all. It would be odd indeed if the standard sexual act turned out to be indistinguishable from a competitive game.

'You have more than satisfied me, Grasshopper, that competitive games are not paradoxical. So let me return to a second objection I promised you I would raise against your theory of constitutive rules.'

'Please do so, Skepticus.'

8 Mountain climbing

In which the Grasshopper defends
his definition by arguing that
some games require a 'limitation in
principle' of the means
a player will permit himself
to use in order to reach his goal

Not all games [I said] are competitive. Therefore not all games exhibit the kind of means-limiting rule which specifies an opponent whose job it is, in effect, to make achievement of the prelusory goal more difficult. There are, that is to say, one-player games. And there accordingly still exists the possibility of discovering or of inventing a game which does not have constitutive rules as you describe them. I suggest that the sport of mountain climbing is such a game.

Sir Edmund Hillary sets out to climb Mount Everest. He will use the best tools available for the job, and although the number and kind he will use are limited, they are certainly not limited by the kind of 'arbitrary' rule that figures in games, but only by how much he can carry. Although a thousand feet of rope would be more useful, he cannot carry that much rope, and although x number of pitons would provide that much more insurance for him, he can carry only x minus n pitons. And so on. He employs *all* the most efficient means available to him. Accordingly, Grasshopper, if you are willing to grant that mountain climbing is a game, you are evidently faced with an example of a game that does not have rules which prohibit more efficient in favour of less efficient means.

I am willing to grant that mountain climbing is a game [Grasshopper replied]. Now, Skepticus, suppose that Sir Edmund, with nearly superhuman nerve and skill, and after escaping death a score of times, has finally arrived at the summit, more dead than alive, but with the nearly superhuman exhilaration that can be produced only by a supreme triumph. As he surveys the panorama of peaks and ridges below him, he is startled to hear himself being addressed in the following words:

'Sir Edmund, I presume.'

Sir Edmund whirls around to see facing him an immaculately groomed Londoner, complete with bowler hat and furled umbrella, and with a copy of that day's *Times* under his arm.

'What are you doing here?' cries Sir Edmund. 'How the devil did you get here?'

'Why, I took the escalator on the other side of the mountain, my dear fellow.'

What would Sir Edmund's response have been if he had known about the escalator beforehand? I suggest that it would have been one or the other of two responses. 1/ He might have decided to ascend the mountain anyway, while adopting a constitutive rule prohibiting use of the escalator. In that case, the resulting instance of mountain climbing would have been a game according to my definition. However, 2/ he might have become completely uninterested in Mount Everest and decided to seek an escalator-less mountain instead. Let us suppose that he acts upon the latter option. Mount Invincible, he finds, is such a mountain, and so he decides to climb it.

Now, what you want to maintain, Skepticus, is that Sir Edmund is trying to play a rule-less game; that is, he is pursuing a goal in such a way that the efforts to achieve it do not depend upon artificially ruling out easier in favour of more difficult means. He is seeking to achieve a state of affairs which is in its natural condition sufficiently challenging. Very well. We find Sir Edmund beginning his preparations to scale Mount Invincible, having made quite sure that no artificial means of ascent have been installed upon its slopes and crags. Before he has progressed very far in these preparations, the bowler-hatted escalator-user meets Sir Edmund at a London club. In the course of conversation he remarks:

'I see by the *Times* that you plan to climb Mount Invincible.'

'That is so,' replies Sir Edmund somewhat coolly.

'Well, there's a beautiful view from the summit. I took a helicopter up there just last week.'

Sir Edmund at once calls off preparations for the ascent of Mount Invincible and begins the search anew. At length he finds Mount Impossible. The most careful testing assures him that the wind currents which perpetually surround the summit prevent a landing by any flying machine or any other kind of mechanical contrivance. The best way, bar none, to get to the top of Mount Impossible is by climbing it, and Sir Edmund climbs it.

Has Sir Edmund succeeded in playing a game with no rules? I think not. It is true that he did not choose some goal, x, and then limit the means he would permit himself to use in achieving it, but he accomplished the same result by doing what he did do. He chose goal x rather than goal y because the means for achieving goal x were more limited than the means for achieving goal y, and the *only* reason he chose x over y was because of that limitation. Therefore, although no overt act prohibiting more efficient in favour of less efficient means was made, that was precisely the effect of choosing the more difficult goal. We may accordingly say, I suggest, that there is here a limitation in principle, for if some new and more efficient means were introduced into the situation (e.g., a flying craft that *could* land), then the available means would once again be insufficiently limited.

There is, that is to say, no difference in principle between creating a challenge by an artificial prohibition of more efficient means to a goal and artificially choosing a goal just because the means for its achievement present a greater challenge than do the means for achieving a different goal. There is no difference in principle between ruling out use of the escalator on Mount Everest and ruling out Mount Everest in favour of Mount Impossible.

But let us put aside for the moment our two additional mountains and return to the real Sir Edmund and the real Mount Everest, where there were no escalators and no flying craft available. Sir Edmund did not, in fact, have to choose between ruling out these devices and selecting another mountain. Everest was fine for his purposes and – you will no doubt wish to contend, Skepticus – he used the most efficient means in climbing it. But suppose we had put to the real Sir Edmund the following question: 'Sir Edmund, there is no escalator to the top of Mount Everest, nor is it the case that anyone is prepared to install one. Still, if that *were* possible (at no expense to yourself, that is understood), would you wish one to be installed so that your ascent would be easier, safer, and more likely to succeed?' It is obvious that Sir Edmund would have said no to such a proposal.

What I have called a limitation in principle (or a subjunctive or counter-factual limitation, if you like) is, it seems to me, necessary in order to explain Sir Edmund's otherwise perplexing response. For it makes clear that Sir Edmund had set himself a lusory goal which required him to *climb* mountains rather than the prelusory goal of simply being at their summits, which would not have required him to climb mountains.

'Well, Grasshopper,' I said when he had concluded, 'you have removed my last doubts about constitutive rules. I would like now to raise an objection of a rather different kind. And I'm afraid that if it is sound it will require a quite radical revision of your definition.'

'Then you must state the objection, Skepticus, come what may.'

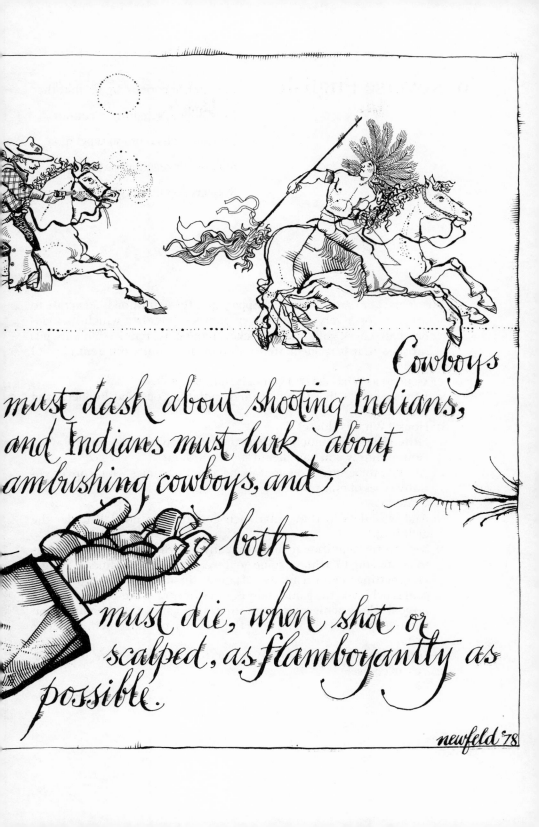

Cowboys
must dash about shooting Indians,
and Indians must lurk about
ambushing cowboys, and

both

must die, when shot or
scalped, as flamboyantly as
possible.

newfeld '78

9 Reverse English

In which Skepticus argues that the Grasshopper's definition cannot capture such common types of make-believe games as Cops and Robbers and Cowboys and Indians

SKEPTICUS: I am satisfied, Grasshopper, that the definition is adequate to account for a very large class of games – the class which includes baseball, chess, golf, bridge, hockey, monopoly, tennis – but I am not satisfied that it is adequate to account for a quite different class of games.

GRASSHOPPER: What class of games is that, Skepticus?

S: I mean games like Cowboys and Indians, Cops and Robbers, and House.

G: House? What is House?

S: Little girls spend much of their time doing what they call 'playing House.' Surely you have heard the expression?

G: Ah, yes, to be sure. You are talking about pastimes which are essentially types of make-believe.

S: Quite so.

G: And why do you think that such activities do not fall under the definition?

S: Because the definition requires that anyone who is playing a game has to be striving to achieve some goal – crossing a finish line, mating a king, getting a certain number of points – in such a way that when the goal is achieved the game ends. But in games of make-believe there is no goal whose achievement terminates the game. Children just go on playing a game of this kind until they tire of it or find something better to do.

G: Still, you would agree, would you not, that these games are activities?

S: Yes, of course.

G: But surely all activities are goal-directed, or at least all intelligent activities (if that is not, in fact, a redundant expression) are. I take it that participation in such pastimes qualifies as intelligent activity?

s: It does.

G: Then surely such activity must have some goal or purpose. Otherwise it would be just a series of random movements.

s: I agree, Grasshopper, that such pastimes have some point to them, that is, some goal. There are, however, two ways in which an activity can be goal-directed. Let us suppose that Jones, who has an hour's wait between trains, decides to 'kill' the time on his hands by playing a game of solitaire. Killing time is his goal and playing solitaire is the means he has adopted to achieve it, for while he is playing solitaire he is in the process of killing time. Now, in order to be playing solitaire, he must be trying to get as many 'up' cards as he can in accordance with the rules of solitaire. That is, the pastime he has chosen as a means for killing time is itself a goal-directed activity. Now consider Jones's daughter, who is with him in the waiting room, and who is also interested in killing time. She decides to play House in order to accomplish that purpose. So she makes believe that she is a mother, and then acts out a number of the things that mothers do. But the motherly things she does are *not* means for achieving some goal analogous to her father's goal of maximizing his number of 'up' cards, for she is not trying to bring about any particular state of affairs. If her father is asked why he makes any given move, his answer will be that it is a means, direct or indirect, for producing 'up' cards. But if she is asked why she does any particular thing, she will reply that that is the way mothers behave, or that that is the kind of thing a mother does. That is, she would refer to a role rather than to a goal. So some activities appear to be goal-governed and other activities appear to be role-governed.

G: You appear to have made something of a *prime facie* case against the definition, Skepticus, I must say, unless, of course, these pastimes are not games at all.

s: Well, they are generally acknowledged to be games by their devotees. 'What game shall we play?' asks young Smith. 'Cops and Robbers,' answers young Jones.

G: I grant that usage must not be ignored in definitional inquiry, Skepticus, even the usage of children. But such usage cannot be finally decisive, can it? Things like Ring Around the Rosie, too, are referred to by small children, by the teachers of small children, and by social scientists who interest themselves in small children, as games. But I think you will agree with me that Ring Around the Rosie is simply a kind of dance to vocal accompaniment, or a choreographed song. It is no more a game than *Swan Lake* is.

s: I agree, Grasshopper, that Ring Around the Rosie and the like are not games, for they are what I should call *scripted* undertakings; that is, activities whose execution is prescribed beforehand, as in a theatrical performance or ceremonial ritual.

G: But are not Cops and Robbers and Cowboys and Indians also, as you say, scripted? Cowboys must dash about shooting Indians, and Indians must lurk about ambushing cowboys, and both must die, when shot or scalped, as flamboyantly as possible. Aren't these things just ritual performances?

s: By no means, Grasshopper. There is, to be sure, something *staged* in these games, but the players are not working to a script. I would say that they were performing a play which had been cast but not written. For the outcome is not known beforehand. Sometimes, for example, the Indians win and sometimes the cowboys.

G: While you were speaking, Skepticus, I have been recalling my childhood, and I must admit that what you say about the enterprise being a kind of casted but unwritten play is quite true. Still, if these things are games they strike me as being highly imperfect games, just as they did when I played them myself. For it was never quite clear what counted as a successful, or even legitimate, move. Young Smith would shout, 'Bang! You're dead, Jones.' And young Jones would respond, as often as not, 'I am not. I ducked in time,' or 'Your gun wasn't even loaded, Smith.' And so on. It was worse than trying to play tennis with imaginary foul lines.

s: Yes, I admit that there is a good deal of that kind of thing in these games. But even if they are rudimentary, or somewhat inchoate, or even partially aborted games, they are still, I believe, in some respect games, and it is that respect that interests me.

G: I wonder, Skepticus, if they aren't merely pretexts or devices for going about shouting 'Bang!' and for 'expiring' in colourful ways.

s: But why do you say 'merely,' Grasshopper? To shout 'Bang!' and to die picturesquely are to play roles, and we have already agreed that these games are essentially role-governed activities.

G: We have *provisionally* agreed to that, Skepticus. But I think we should be very cautious in giving the thesis our unqualified assent, for if there are two radically different kinds of game – role-governed and goal-governed – then we would have to give up our attempt to formulate a single definition of games.

s: Unless, of course, we could come up with a more general definition which would satisfactorily account for both kinds.

G: Well, yes, to be sure.

s: I have one.

G: I beg your pardon?

s: I have a definition which gives an adequate account of both goal-governed and role-governed games.

G: Indeed.

s: Yes. It supersedes your definition, which is adequate to account only for goal-governed games.

G: (mutter, mutter)

s: I beg your pardon?

G: I said nothing, Skepticus. Pray expound your definition.

s: Very well. Simply put, it is that games reverse the ends and means of other activities.

G: Perhaps you could amplify that a bit.

s: Certainly. The idea was suggested to me by Kierkegaard, for in his 'Diary of a Seducer' the diarist makes a kind of game out of a love affair precisely by means of such a reversal. Whereas a serious seducer plots and plans so that he can achieve what I suppose we may call *habeas corpus*, Kierkegaard's diarist adopts *habeas corpus* as his goal only so that he can be plotting and planning to achieve it.

G: Yes, Skepticus, now that you remind me, that is precisely what Kierkegaard's diarist does. Just that kind of switch performed on ordinary activities is what Kierkegaard calls the 'aesthetic' treatment of life and is a cardinal principle in what he archly calls a 'theory of social prudence.' The idea, in somewhat different form and with a different application, first appears, I believe, in Kant's *Critique of Aesthetic Judgment*, where Kant likens aesthetic experience to play as a kind of 'purposiveness without purpose.' The idea can also be found, along with many others with which it is almost hopelessly entangled, in the rather swampy dialectic of Friedrich Schiller's *On the Aesthetic Education of Man*. And I believe the sociologist Georg Simmel expresses very much the same kind of notion when he observes: 'This complete turnover, from the determination of the forms by the materials of life to the determination of its materials by forms that have become supreme values, is perhaps most extensively at work in the numerous phenomena that we lump together under the category of play.'* It is true that these writers were addressing them-selves primarily to 'play' rather than to games, but since none of them made an important distinction between playing and playing games, I think you are justified, Skepticus, in treating the idea as an idea about games.

s: (*with some testiness at being thus upstaged*) I think we can skip these questions of provenance and affiliation, Grasshopper. That some-thing like this idea was first expressed by Kant is hardly the issue.

G: To paraphrase Sir Winston Churchill's remark about the function of the cavalry in modern warfare, reference to eminent figures of the past serves to lend tone to what would otherwise be merely an honest search for the truth.

s: *May* I get on with it, Grasshopper?

G: By all means, Skepticus.

s: Make-believe, I suggest, is a kind of impersonation. But whereas what might be called *serious* impersonators play roles so that they will be taken for the subject of the impersonation, in make-believe the per-formers take a subject for impersonation so that they can be playing the roles such impersonation requires. An impostor behaves like a Russian princess in order to be taken for Anastasia, but a player at make-believe chooses to impersonate Anastasia so that she can be-have like a Russian princess.

G: You seem to be saying that people who play at make-believe put a kind of reverse English on life's genuine enterprises.

s: Reverse English?

G: Yes, for we may say – not too fancifully, I think – that the governing purpose of, for example, an ordinary billiard ball is to depart from the point of impact, whereas the tendency of an Anglicized billiard ball (as I suppose we may call it) is to *return* to the point of impact. Its departure is not its final purpose but a preliminary condition neces-sary for its subsequent return. Similarly, the purpose of a genuine imposter in playing a role is to produce a false identity, while a player at make-believe assumes a false identity so that he can be playing a role.

---

* More specifically, the four sources cited here by the Grasshopper are as follows: Kier-kegaard *Either/Or* part 1 'The Rotation Method' and 'Diary of a Seducer'; Kant *The Critique of Aesthetic Judgment* first book, chapters 9–11; Schiller *Letters on the Aesthetic Education of Man*, especially letters 14, 15, and 26; Simmel *The Sociology of Georg Simmel* edited by Kurt Wolff (The Free Press 1950) 42.

s: Precisely, Grasshopper, precisely.

G: Well, Skepticus, that is an interesting way to look at make-believe, I must admit, but it is not clear to me that reverse English can also account for the *goal*-governed games that we have been considering until now.

s: Oh, but it can, Grasshopper. Do you recall that you earlier used the example of high-jumping to illustrate the original definition?

G: Yes, I remember. I used the example to show that games involve a limitation of means, since high-jumpers intentionally place obstacles in their own paths.

s: Quite so. But notice that commitment to such an enterprise involves reverse English every bit as much as does a commitment to make-believe. For a genuine surmounter of obstacles does so in order to get to the other side, but a high-jumper tries to get to the other side only so that he can be surmounting obstacles. High-jumpers and players at make-believe are both playing games by putting reverse English on some serious pursuit. The only difference between them is that one kind of game calls forth dramatic skill and the other kind calls forth athletic skill.

G: Again, Skepticus, I find your suggestion plausible, and although I have one or two reservations about it that I would like to put to you in due course, let me applaud your identification of dramatic ability as the skill appropriate to a distinct class of games. For if that fact were more widely recognized, such recognition might result in a much needed corrective of our lusory institutions as they now exist.

s: What do you mean?

G: Well, as we both realized when we began talking about make-believe, Skepticus, games of this kind are nearly always played by small children, and as played by small children they display rather serious defects. Goals, rules, strategies – all appear unclear and unfixed. And often such enterprises seem to be less games than dramatic projections of day dreams or fantasies. And so they are soon abandoned in favour of the unambiguous games that have succeeded in becoming established institutions: athletic games, board games, card games, and so on. Dramatic skill continues to exist in only the most attenuated form in parlour games like Charades, where it is very strictly subordinated to the arts of puzzle solving and coded communication. But I suspect that there is nothing about dramatic skill which makes it inherently unsuited to being the chief, rather than a severely subordinated, element of well-constructed games. If so, that fact could have some fairly important practical implications. Everyone is familiar with the practice of sending teen-age boys outside to do something athletic when their surplus energy turns to horseplay and begins to endanger the furniture and their younger siblings. But other people – adult as well as adolescent – can be just as annoying or destructive with their dramatic horseplay. So it might be quite useful to have a game outlet for people who are always starting unnecessary arguments or reacting histrionically to imagined affronts or invented crises just because they are bursting with dramatic potentiality. Make-believe pastimes seem to provide such outlets for children, and if such pastimes are indeed games, we ought to find out how they work, so that they can be improved and instituted as socially acceptable adult pursuits. I am thus keenly interested in your suggestion that the make-believe pastimes of children are rudimentary games, even though I am less than convinced that reverse English is what makes them games.

s: Perhaps I can convince you that that is so by means of the following illustrative tale.

G: Perhaps you can, Skepticus. I am certainly willing to listen.

He realized that he had no interest in the military, or even, he had to admit in the patriotic value of his assign ments, but only in the opportuni ties they afforded him for performing dramatic roles.

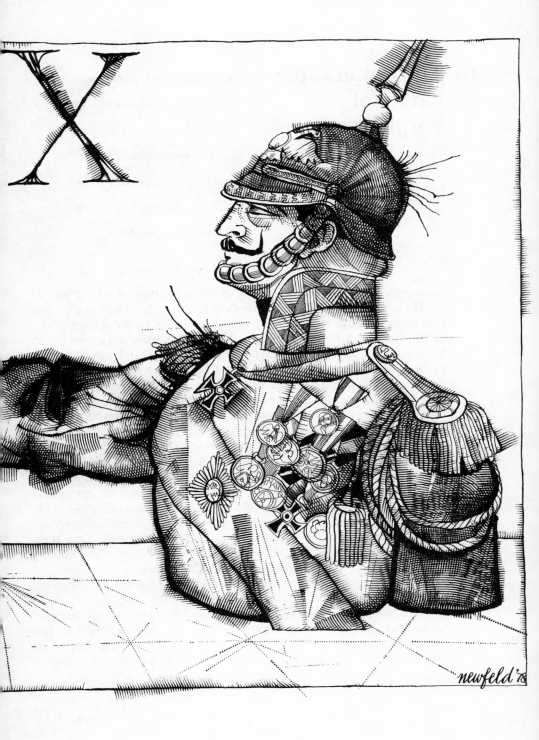

newfeld '78

10 The remarkable career of Porphyryo Sneak

In which Skepticus pursues the attack he began in Chapter Nine with a tale of espionage and impersonation, and the Grasshopper begins a counter-attack by extending the tale Skepticus had begun

Porphyryo Sneak [I began] was the last and greatest of a long line of Sneaks who through six generations had brought the arts of impersonation and espionage to a state of virtual perfection in the service of the British Crown. And just as, in that kingdom's earlier days, one might have said of one of its monarchs, with reverence and awe, 'He is a Plantagenet and a king!' so the letter of introduction which young Porphyryo brought with him to Secret Service headquarters in 1914 contained, and needed to contain, only one sentence in order to ensure his immediate employment: 'Bearer is a Sneak and an imposter.'

The young Sneak's first assignment was to impersonate General Kriegschmerz, a battle-weary member of the German High Command who had secretly defected to England. As Kriegschmerz, Sneak was able without difficulty to obtain valuable strategic information and return it safely to England. Upon his return, however, there was no immediate need for his services, and he quickly became bored and depressed. This depression lasted until his next assignment, when he immediately regained his customary cheerfulness. The alternation of these moods then became the pattern of his life. He felt really alive only when he was playing a part, and the intervening periods were merely empty times of waiting to be called on stage. That was the first phase of Sneak's incredible career.

Then one day Sneak made an astonishing discovery. He realized that he had no interest in the military, or even, he had to admit, in the patriotic value of his assignments, but only in the opportunities they afforded him for performing dramatic roles. With this new information about himself he adopted a quite new attitude towards the conduct of his life. For he saw that he need not simply sit around waiting for an assignment to be handed out to him. He could, instead, seek out such assignments. And so Sneak became a double agent; not, of course, to double his income, but to double his roles. And during good times – the second world war, Korea, Vietnam – Sneak became a triple, quadruple, and quintuple agent, and so also became perforce not only the world's most versatile character actor, but also the world's quickest quick-change artist.

The next stage of Sneak's career began when a sudden and unexpected decrease in international tensions resulted in a sharp reduction in the military (and espionage) budgets of all the nations of the world. There was, accordingly, scarcely any demand for Sneak's services, and he soon found himself worse off than he had been in the old Kriegschmerz days. Sneak, as might be expected, did not waste time in idle despair, but at once took steps to remedy the situation. Thanks to the vast amount of intelligence he had amassed in the course of a career which had already included impersonating all the important heads of state of the world, together with most of their cabinet ministers, Sneak was in an excellent position to re-create all the suspicions, jealousies, and fears that had hitherto characterized the intercourse of nations, so that once more immense funds were allocated to the espionage establishments of the world. Sneak's services were now more in demand than ever, and he was once again a happy man, indeed doubly so. For not only had he refashioned events so that opportunities for his dramatic fulfilment were again at a maximum, but he was secure in the knowledge that if ever international peace and goodwill threatened again to overtake the affairs of the world, he was in a position to restore matters to a more satisfactory condition. And that is just what happened in this period of Sneak's life. When tensions began to relax and espionage budgets were cut, Sneak would assume an appropriate role and, with a word here and a frown there, thicken up the plot again.

Things continued in this way for several years, until Sneak made his next great discovery. He had always been more or less aware of the fact that his life was made up of two distinct kinds of enterprise: on the one hand dramatic acting, which was the ruling passion of his life, and, on the other hand, the things Sneak had to do, or put up with, as undesirable but necessary conditions for his being able to satisfy that passion. First the necessary evil had been simply the boredom of waiting in the spy's squad room to be called up for duty. Then, when he became a multiple agent, it had been making out tedious job applications and undergoing idiotic interviews in Viennese Ferris wheels, Bessarabian brothels, and Levantine latrines. And now, on those occasions when Sneak found it necessary to take a direct hand in international politics, it was the necessity of assuming a role so that he could sow and cultivate the seeds of global dissension. At this point Sneak paused in his reflections and re-examined that last thought, for there was something odd about it. Then it came to him. Somehow it had come to a point where the necessary inconvenience he had to accept as a condition for future opportunities for dramatic acting was itself dramatic acting. But in that case, of course, it was not an inconvenience at all; it had become part of the game. Sneak congratulated himself on his good fortune.

With this new realization Sneak also realized that to employ his skills at political contrivance as an intermittent corrective of the dramatic defects attendant upon international peace and goodwill was not the most efficient way to go about the business. Since his political manœuvrings and the opportunities for espionage which it was their purpose to maximize both took the form of dramatic impersonations, there was no reason why Sneak should not take a continuing, rather than a merely remedial, hand in world history. Accordingly, his impostures began to be governed as much by decision-making as by intelligence-gathering considerations, a fact which was to have far-reaching consequences for his future. For he next discovered, to the delight of his dissembling soul, that these purposes could both be accomplished in the course of performing one and the same role. Thus, during that bitter January of 19__ Sneak was on assignment in Ottawa for the CIA. His orders were either to confirm or to deny the existence of a suspected secret military alliance between Canada and the Soviet Union. In order to accomplish his mission Sneak slipped into a Gallic shrug and appeared in Ottawa as the Canadian prime minister. He was easily able to obtain the required information. There was, in fact, *no* such alliance, and therefore no reason whatever for uneasiness on the part of the United States. This was, of course, bad news for the spy business. Therefore, Sneak, still in the role of prime minister, issued a public statement denying that Canada's military treaty with the USSR was in any way indicative of strained relations between Canada and its good friend to the south. And next week on assignment in Moscow Sneak took the opportunity as Soviet premier to issue a strong warning to the People's Republic of

China. The spy business boomed, and Sneak was secure in the knowledge that he was in a position (as Kierkegaard said of people like his seducer) to play at battledore and shuttlecock with the whole of existence.

GRASSHOPPER: Well, Skepticus, that is an amazing story, to be sure. But I wonder if it is quite the whole story?
SKEPTICUS: What do you mean?
G: Your reference to battledore and shuttlecock evoked in me a kind of vision of Sneak's future. Perhaps you would like me to tell you what I saw.
S: By all means, Grasshopper.
G: Very well. The rest of the story, then, goes like this.

Happy though Sneak was in this felicitous arrangement of his and the world's affairs, there was yet another revelation in store for him. It came about in the following way. In April of 19— Sneak was impersonating the Queen of England on assignment for Swiss counter-intelligence. Normally Sneak would not have accepted an assignment from a client so undistinguished in espionage circles, but he had a special reason for wishing to be the Queen just then. For the previous day he had, as Eggbeat of Nog (that is, as head of state of the principality of Nog), issued an official note to the Crown requesting the admission of Nog to the British Commonwealth of Nations. And now, because of some rather complex plans in another area, Sneak was most interested in seeing to it that England, for the moment, give neither an affirmative nor a negative response to the Eggbeat's request. As Queen of England, therefore, he stated publicly that Her Majesty's ministers and advisers would take the request under advisement for a fortnight. Then Sneak sat back to await results. And while he was waiting the final revelation came to him. *He had just made a counter-move to his own move.*

And it was at precisely this moment that Sneak came in from the cold. His brilliant career as a spy was over, and he entered upon the next stage of his career with an intoxicating sense of freedom. For he realized that he could play any role in any dramatic situation he chose to contrive, quite independently of the demands, direct or indirect, of the spy business. He was his own playwright, and a kind of God, for the whole world had become his stage.

The brief period of modern history which resulted from Sneak's great discovery became known as the Mad Months. And it seemed, during May, June, and July of 19__, that the entire fabric of international relations was simply shredding to bits. Alliances between nations were formed and dissolved with dizzying speed, cabinets were reshuffled daily, and the world suffered continuous vertigo as it peered in terror over one brink of disaster after another.

Fortunately for human civilization, Sneak was not a God but a mortal man. While he had begun his latest exploits by thinking of himself as a kind of omnipotent actor-writer-producer, he soon began to see himself as trapped in an interlocking series of hectic one-man badminton games where, just to keep the bird in play, he had not only to be running from one end of the court to the other, but also from one court to another in an endless line of courts, until it seemed that the whole of existence was playing at battledore and shuttlecock with *him*. Inevitably it was too much. He suffered a nervous collapse, and the world returned to a more tolerable level of catastrophe and recovery.

Sneak, meanwhile, had the good sense to get himself admitted to a reputable sanatorium, where he received expert medical and psychiatric care. The following dialogue is a verbatim account of his final therapeutic session.

DR HEUSCHRECKE: Please have a chair, Mr Sneak. No, not the couch, please. I am not a psychiatrist.

SNEAK: Well, that makes a nice change, at any rate. What are you, then?

H: A doctor of philosophy.

S: Oh, a PH D.

H: No, Mr Sneak. To put it more accurately, I am a *physician* of philosophy. I try to cure the philosophical maladies of my patients. You have been sent to me because my psychiatric colleagues have been able to find nothing whatever wrong with you psychologically.

S: But what about my breakdown?

H: Sheer physical exhaustion.

S: Yes, but that exhaustion was brought on by some deep-seated psychic disturbance, was it not?

H: I'm sorry to have to disappoint you, my friend, but it was not. It was brought on by overwork.

S: But if that is so, why am I in such a depressed state? I thought my condition had been diagnosed by your colleagues as melancholia.

H: That was an early provisional diagnosis, to be sure, but it has proved to be incorrect. You are not a melancholic. You are simply melancholy.

s: Do you mean to say there is nothing seriously wrong with me?

H: I mean to say there is nothing *clinically* wrong with you at all.

s: Then there *is* something wrong with me?

H: There is.

s: Tell me what it is, Dr Heuschrecke, for God's sake!

H: You are suffering from a logical fallacy.

s: A logical fallacy! What on earth do you mean?

H: I shall try to explain it to you. We will have to begin by going back to your childhood.

s: My childhood? I thought you said you were not a psychiatrist.

H: My dear fellow, you mustn't think that psychiatrists have a monopoly on childhood.

s: Oh. Sorry.

H: Your childhood was occupied to an abnormally large extent with make-believe. In fact, you were something of a prodigy at this pastime, going far beyond the usual childhood games of Cops and Robbers and the like. Actresses and Bishops was one of your early inventions, and this was quickly followed by others: Lawyers and Clients, Priests and Confessors, Princes and Parliaments, and Presidents and Impeachers, to name just a few.

s: (*relaxed and smiling now*) Yes, those were happy times.

H: Just so. And it was thus quite natural that when the first world war came along you should be attracted to that branch of military service where you could serve your country best by doing what you relished most. Then (and I realize I am not telling you anything you do not already know), during those periods of idleness between assignments, you came to realize that impersonation was not, for you, primarily a means for serving your country, but that the kind of service to your country which you were best able to provide was a means for you to be engaged in impersonation.

s: Quite right.

H: I would like to suggest to you that, even this early in your career, espionage was very much like a game for you.

s: That describes very well my attitude towards my profession.

H: Yes, and this is further supported by the fact that once you had made this discovery about your attitude, you forthwith became a double agent. If patriotic goals were in fact merely devices which enabled you to perform dramatic roles, then there was no reason why you should not increase the frequency of those roles by providing your services to more than one *patria*.

s: Quite true.

H: And the same kind of reverse English – if I may put it that way – explains your next step, which was to ensure opportunities for espionage by keeping international relations in a state of ferment. And this goal of exacerbating world tensions you accomplished by role playing. So now you had two theatres, so to speak, for your dramatic performances, and you congratulated yourself on this happy turn of events.

s: Quite right.

H: Now, throughout the progress of your career up to this point the governing end which the application of reverse English had turned into a means was deception, was it not? That is, your goal of producing dupes was, directly or indirectly, really a means which enabled you to perform roles. Normally one impersonates so that one can produce a dupe, but you sought to produce a dupe so that you could impersonate. This takes us to the crucial turning point of your life, the Eggbeat affair. Just as, when you became a double agent, you eliminated patriotism as your pretext for duplicity, in the Eggbeat affair you eliminated duplicity as a pretext for impersonation. For in that affair no dupe was produced. What was produced was the opportunity for further role-playing by the responses of a make-believe Queen to a make-believe Eggbeat. When you realized this you felt an intoxicating euphoria, as though you had at last been released from heavy chains and, as is your wont, you immediately acted on the basis of that realization. You very sensibly came in from the cold, and you then, much less sensibly, embarked upon that course of events which resulted in the Mad Months.

s: Everything you say is quite true, Heuschrecke, but what is the logical error you claim lies at the bottom of my problem, and how will its correction bring about my rehabilitation?

H: Your error is the same as that of the fabled inventor of roast pig. And just as his error was correctable, so is yours. You remember the story. One day this chap's barn burned down, killing a pig he kept there. Finding the flesh of the burnt pig palatable, indeed delicious – having, that is, invented the pork roast – he sensibly decided to avail himself of more roast pork on future occasions. So he rebuilt his barn, put a pig in it, and set fire to the barn, thus committing arson and a logical fallacy.

s: The fallacy of mistaking a sufficient for a necessary condition.

H: Precisely. What you wanted were opportunities for playing make-believe games, and you found – by accident, just as the pork fancier had – that impersonating monarchs, prime ministers, and presidents provided such opportunities. Then, again like your predecessor in fallacy, you mistakenly supposed that such impersonations were not merely sufficient for your dramatic purposes but also necessary. And just as we may imagine that the pig chap came to financial ruin by having continuously to rebuild his barn, we observe that you very nearly destroyed your health by the excessive expenditures of energy required to keep your global badminton tournament in progress. And if we add to the pig fable the embellishment that each time the primitive gourmet burnt his barn the whole community was threatened with incendiary destruction, we have a complete parallel to your own case. And we also have an indication of where the solution lies, do we not?

s: Well, I suppose the pig chap corrected his logical error by inventing the cook stove.

H: The cook stove, precisely. And the Eggbeat affair shows us quite clearly where to look for your own particular cook stove, does it not?

s: It does?

H: Of course it does. Disaster overtook you only because of the sheer size of the arena you supposed you needed for your dramatic performances. For the Eggbeat affair prompted you to make the whole world your stage and all of the world's roles your personal repertory. But the Eggbeat affair holds a more profound revelation which you did not give yourself time to fathom.

s: What is that, Heuschrecke?

H: *It is, Sneak, that you had spent a life-time in discovering that what you liked to do best in the world was to play make-believe games.*

s: Good Lord.

H: Yes. For with the Eggbeat affair you were not spying, you were not creating a crisis for the sake of the espionage business, you were not duping someone so that you could be playing a role, you were not even duping someone into making a response which would enable you to make an answering response. You were not, that is, doing anything which was merely an enabling manœuvre or merely a pretext for playing a part. All your moves were playing a part and nothing but playing a part. You were not engaged in any kind of imposture, although you thought you were.

s: What was I doing, then?

H: You were playing Heads of State. And you hadn't had so much fun since those nearly forgotten days of Actresses and Bishops. If you will forgive a fairly revolting but nonetheless apposite observation, you had travelled the whole world over seeking the bluebird of happiness only to find it in your own back yard.

s: Oh, come now, Heuschrecke, if that is so, then I could have played Eggbeat/Queen without stirring out of my own living room.

H: Precisely.

s: (after an appreciable pause) You are saying that that is my cook stove.

H: I am.

s: I don't know, Heuschrecke.

H: What don't you know?

s: Well, you seem to be telling me that my rehabilitation will consist in my sitting in the parlour talking to myself.

H: But that's exactly what you were doing in the Eggbeat affair – talking to yourself.

s: Yes, but good heavens, Heuschrecke!

H: What is it, Sneak?

s: Earlier, when you were saying that the Eggbeat affair was really a game of Heads of State, and that I had recaptured my childhood by playing it, you meant that literally, I take it.

H: Yes, I did.

s: And you really are suggesting that I spend the rest of my life playing childish games?

H: I'm certainly suggesting that you spend the rest of your life playing games. Whether they are childish or not depends on the games you choose to play, doesn't it? I wouldn't expect you to play Cowboys and Indians or Cops and Robbers, or even Actresses and Bishops. But then I wouldn't expect Bobby Fischer to spend the rest of his life playing checkers either, although I am quite sure that he will spend the rest of his life playing games. But what do you find so repugnant about the idea of playing games for the rest of your life? That is all you have been doing with your life so far.

S: Yes, yes, I do see that, Heuschrecke, but you have to admit that there is no small difference between impersonating the Queen at Buckingham Palace and playing Heads of State in my living room.

H: Of course there is a difference, there is a tremendous difference. But the question is whether that difference makes any difference to *you*.

S: Well, it would have to, wouldn't it?

H: No, it wouldn't. Furthermore, I don't believe it does.

S: Then why am I carrying on so about it?

H: Because you are playing a role appropriate to the occasion. You couldn't resist such a golden opportunity.

S: (*throws his head back and laughs*) You're right, of course, but how did you see through me? Am I slipping?

H: Your acting was flawless, as a matter of fact.

S: Then how?

H: Because I trusted the psychiatrist's reports on you.

S: What does that have to do with it?

H: They established conclusively that your sole motive in playing what we ought now to call the games of Espionage, World Crisis, and Heads of State was entirely a *game* motive. If they had established the fact that playing these games were devices to serve other (probably neurotic) purposes, then I would not be at all confident of your rehabilitation along the lines I have suggested. Thus, if you had a kind of compulsion to be deceiving people, then playing pure make-believe – that is, make-believe with all of your cards on the table – would not, of course, meet your requirements. But your reaction to the Eggbeat affair was sufficient to rule that motive out of the picture. For it was then that you realized that you could engage in role playing even more effectively *without* practising deception. But if deception had been your motive, you would not have come in from the cold when you made this discovery; you would have gone back to full-time spying. It was that fact which put the psychiatrists on the right track. Having eliminated the compulsion to deceive as the ulterior purpose or hidden cause of your game playing, they next considered all the other possible motives and causes they could think of: exhibitionism, polymorphous transvestism, pernicious misanthropy, generalized social disgust, nagging birth trauma, aggravated atavistic rage – you know, the lot. But they couldn't pin a thing on you. They wisely, and I must say surprisingly, concluded that your problem was not psychological but logical, and so they sent you along to me.

s: The psychiatrists must have found me something of an anomaly.

h: Oh, unquestionably. It is, after all, wildly improbable that there should exist a person who has done the things you have done solely because he has made a mistake in logic. You seem, even to me, much less a real person than something invented to illustrate a principle in a treatise on the philosophy of games.

s: Ha-ha.

h: But you are a real person, and my patient, and there is more to be said about your rehabilitation, so let us continue. We had got to the point, you will recall, where you were expressing mock dismay at a future which appeared to consist in babbling to yourself. But of course you need not confine your future activities to make-believe on the model of the 'Eggbeat'/'Queen' game. You need not, that is to say, play only solitaire make-believe. Notice, first, that make-believe is normally a two-role game, even though both roles may be played by the same person, as in the 'Eggbeat' instance of Heads of State. Thus make-believe is not the same as mimicry – that is, impersonation as an end in itself. If it were, then your rehabilitation might consist in going on the stage as a master impersonator. And notice also that make-believe is not the same as playing a part in a stage play. If it were, your rehabilitation would of course be accomplished by your becoming, in all likelihood, the greatest theatrical actor in the world. How would that strike you?

s: Not at all well. Acting out a part in a play is simply being enslaved to some script writer. It is like miming the moves in a game which has already been played by someone else.

h: That is what I thought you would say. And it points to a basic feature of make-believe games. Each 'move' (if we may call it that) either is for the purpose of evoking a dramatic response, or is such a response, or is both. But these evocations and responses really are evocations and responses; they are not merely representations of such interplay, as is the case in staged performances. The players are, in a way, writing a script at the same time that they are enacting it.

s: Quite right.

h: Now, looked at from the viewpoint of one of the players in a two-role game, what he wants the person performing the other role to do is to keep providing him with opportunities for dramatic responses (e.g., feeding him 'good' lines). There are two ways in which a player can achieve these results. His 'partner' in the game might provide such opportunities because he is also a player in the game or because he has some other reason for providing such opportunities. Among non-players, the providing of such opportunities might be quite unintentional, and this was the case with all of the games you played prior to "Eggbeat." By deceiving your 'partners' about your identity in these games, you caused them to give you lines (and the like) that suited your dramatic purposes, even though it was not their intention to be suiting those purposes. You were playing a two-role, two-person, one-player game. It is also possible to play a two-role, two-person, one-player game where the person who is not a player (but, in effect, a device) provides you with dramatic opportunities with the conscious purpose of doing so. Instead of having to dupe someone into performing the enabling service, you forthrightly ask him to do so. You might even offer an inducement for his service, such as a cash payment. This is just what Gamma Rex in Gilbert and Sullivan's *Princess Ida* did, even though the mercenaries he hired botched the job:

> I offered gold
> In sums untold
> To all who'd contradict me –
> I said I'd pay
> A pound a day
> To anyone who kicked me –
> I bribed with toys
> Great vulgar boys
> To utter something spiteful,
> But bless you, no!
> They *would* be so
> Confoundedly poli*te*ful!
> In short, these aggravating lads,
> They tickle my tastes, they feed my fads,
> They give me this and they give me that,
> And I've nothing whatever to grumble at!

But the best way to get good lines is for your partner to be a player, because then he has a motive which is better than that of either of the others. The dupe is worst, of course, because he is least dependable, and most of the time he isn't giving you lines at all but going about his own affairs. And the person who feeds you lines for some reward (or out of friendship or fear, it might be added), although we would expect him to be more constantly employed at his task than the dupe, is only indirectly motivated to provide the desired service. Only another player (or yourself as the other player) has a direct motive. For he must give you good lines in order to get good lines in return, and since you are motivated to do the same for him, the game is itself a reciprocating system of role-performance maximization.

s: And you are telling me that that, in principle, is the kind of game that I have spent my life in playing?

h: I am.

At this point I could contain myself no longer. 'But, Grasshopper!' I exclaimed, 'Heuschrecke's description of Sneak's games contains no reference to the fact that the roles performed in them must be *assumed* roles, and so it misses the very essence of what Sneak was up to; namely, the application of reverse English to genuine imposture.'

'Yes, Skepticus,' he replied, 'I am aware of that omission, and I find its absence from Heuschrecke's definition very suggestive indeed.'

'Oh, it is *suggestive* enough,' I replied with some heat, 'for it suggests that reverse English and assumed roles have nothing essentially to do with the kind of game we have been trying to define.'

'Precisely,' said the Grasshopper, 'and that has been my suspicion from the beginning.'

'Well, it has not been my suspicion and it is not my suspicion now. And I don't think Sneak, who surely knows better, should let Heuschrecke's definition go unchallenged.'

'As a matter of fact, he doesn't, so let us follow their colloquy a bit further.'

'I should think so,' I replied.

s: You have nearly persuaded me, Heuschrecke. There is just one thing that bothers me about your description of my game as a reciprocating system of role-performance maximization.

h: What is that?

s: The description makes no reference at all to the fact that in make-believe one *assumes* a role which is not the player's real-life role. That strikes me as a rather startling omission.

H: On the contrary, I don't think it's an omission at all.

S: But surely playing a part is the very essence of make-believe.

H: Playing a part is, yes. But playing what might be called a foreign or assumed part is not. One can also play, so to speak, native or proprietary parts.

S: What on earth is a proprietary part?

H: One way to define it is as follows: a part of such a kind that when one plays it, one is not conveying misinformation about one's identity. If, for example, I have the job of lookout for a band of bank robbers, and if I want to give myself a plausible reason for loitering in the vicinity of the bank, I might (taking advantage of my short stature and youthful appearance) put on a boy scout uniform and help old ladies across the street. I take it that you would accept this as an example of someone playing a part which is not his own part.

S: Obviously.

H: Very well, now suppose that a boy scout does the same thing. He dons his uniform and helps old ladies across the streeet. He is also playing a part, but it is his own part; that is, its performance conveys information rather than misinformation about the performer. But – and this is the point – the part itself is just the same in the two cases.

S: You are talking about role-playing in everyday life.

H: Precisely.

S: You sound like a sociologist.

H: That can't be helped. The point is that there are roles which enjoy a kind of objective or public status, so that they can be performed by different people for different purposes. They are in this respect like clothing. All kinds of apparel are for public sale, and I can purchase and put on something which correctly conveys my position in life, or I can purchase and put on something which misrepresents my position in life. For example, I can put on a business suit or I can put on the uniform of a full admiral. The only difference is that suits and uniforms are patterns of cloth and roles are patterns of behaviour.

S: Yes, well, I'll concede that what you say is highly plausible, but I don't see what that has to do with the problem before us. Even if I grant the distinction between proprietary and assumed roles it still seems clear to me that make-believe must necessarily consist in the performance of assumed roles.

H: I agree that what you say seems intuitively obvious. It is, nevertheless, untrue. But I see that our time is up for today. We will have one last meeting tomorrow just before your discharge, but overnight I would like you to read this (*Heuschrecke produces a manila folder from a desk drawer*) and then bring it back with you tomorrow.

s: (*taking it from Heuschrecke's hand*) What is it?

H: It is the case history of another patient.

s: You want me to read someone's confidential file?

H: It's perfectly all right. I have his permission.

s: Very well, then, I'll be glad to. Till tomorrow, then.

XI

You are suffering from a logical fallacy.

An episode in Drag's early life [Sneak read] set the pattern for everything that was to follow. As a boy scout young Bartholomew was interested almost exclusively in those scouting practices which fall into the good deeds department, and in that department he was especially keen on the good deed which consists in helping old ladies across the street. After a bit Bartholomew came to value the role of Old Lady Helper (Streetwise) at least as much as he did the benefit which performance of that role is presumed to provide for old ladies. He thus took to lurking about the busy intersections of the city where he lived which, luckily for him, was St Petersburg, Florida, so that Bartholomew enjoyed a veritable glut of opportunities for performing his service and, more important, his role. But one traumatic day his family moved from St Petersburg to Doze, a hamlet in the hinterland of the state where the entire female population was under the age of forty-five. This was the worst calamity that had ever befallen Bartholomew in his young life, and it might have been too much to bear except for one saving development. It was decided that Bartholomew's elderly grandmother should come to live with the Drag family in their new house.

After the family's removal to Doze Bartholomew's efforts became concentrated upon getting his grandmother to wish to cross the town's one street. The various artifices Bartholomew employed in accomplishing these arrangements need not concern us in this report. Suffice it to say that Bartholomew was now engaged in playing a two-role, two-person, one-player game of the kind in which the non-playing participant is manœuvred into performing the desired complementary role. But grandmothers are much less easily deceived by small boys than small boys believe, and Grandmother Drag very soon realized what her grandson was up to. She was, however, an especially indulgent grandmother, and so she was willing to humour Bartholomew in his pastime. And so at this point Bartholomew was playing a two-role, two-person, one-player game where the non-playing participant intentionally performed the complementary role out of, we may say, the goodness of her heart. But, as often happens in dealing with small boys, this favour was exploited rather than returned, and Bartholomew, since his appetite for the game knew no bounds, very soon became a dreadful bore and nuisance to his grandmother.

And so she became much less available for sorties across the street to the public library, the post office, or the candy store. Whereupon Bartholomew quickly restored affairs to their original satisfactory condition by producing a bribe. It was unmistakably conveyed to Grandmother Drag that Bartholomew's usual sunny disposition would be replaced by an attitude of sullen bad temper if the grandmotherly excursions fell below a certain level of frequency.

The rest of Drag's life, in those particulars which are relevant to his treatment and rehabilitation, consisted of a series of arrangements which were in essence the same kind of arrangement he had achieved between himself and his grandmother.

By the time Drag was thirty-five he had accumulated, as all of us do, a quite extensive repertory of proprietary roles; all the roles, that is, associated with the various social positions he occupied: father, husband, boss (he was owner-director of a computer manufacturing corporation), chairman of the Opera Board and of the Heart Fund, and city councilman, to name just a few of his more obvious positions. And since each of these and similar positions has a number of distinct roles associated with it, Drag was, like the rest of us, called upon to perform many different roles in the ordinary course of events. And many of them he performed, as the rest of us do, largely automatically and unreflectingly. But with respect to a very substantial number of them Drag assumed a distincly atypical posture. He treated them just as he had long ago treated the role of helping old ladies across the street. That is to say, he valued performing them at least as much as he valued their social benefits. Among his favourites were: Understanding Father, Understanding Husband, Pig-Headed Father, Pig-Headed Husband, Graciously Condescending Banterer (Typing Pool), Ditto (Assembly Line), Jocular Chairman of the Board, Gruff Chairman of the Board, Sympathetic Confidant, Shocked Confidant, and many others. And since Drag valued these roles not primarily for their social uses but as vehicles for dramatic performance, there developed a hiatus between Drag's performance of the roles and the situations in which they could appropriately be performed. That is, he took to performing them even when the situation did not require it, just as he had done as a boy scout. And the other people who happened to be involved in his performances were treated in the way that he had treated his grandmother, that is, as dramatically enabling devices. And at first, just as was the case with his grandmother, the other members of the cast in his little dramas were unintentional and unknowing accomplices. But it soon became clear to them what Drag was doing. Now Drag was an immensely likable man, and his friends and acquaintances, when they realized that Drag had a kind of quirk, were entirely ready to humour him in what they were prepared to accept as a minor peculiarity in his make-up. But, as it had been with Grandmother Drag, the more they humoured him by pandering to his eccentricity, the more demanding Drag became of their services. He became, in short, a nuisance and a bore. A senior stenographer in the typing pool would whisper to a junior typist, 'Go over to the water cooler and banter with Drag,' or a husband would say to his wife, 'I've got to think up some personal problem I can confide to Bart on the golf course tomorrow or he'll be grumpy all day.' Or: 'Smith, I'm going to have to muck up your figures on this Jessup Corporation order. I'm sorry, but we'd better give the old man the opportunity to hit the ceiling tomorrow.'

Finally things reached a crisis stage with the Robinson affair.

'Robinson won't get his promotion, you know.'

'Why not?'

'The old man kept him in his office till midnight last night playing Indecisive Executive over the Kramer account. Finally Robinson got fed up and said he had better things to do than play parlour games all night.'

'Oh Jesus. What did the old man do?'

'Oh, he just whipped out Understanding Boss in the Face of Extreme Provocation and apologized to Robinson for keeping him so late.'

'Look, we can't let this kind of thing happen again. Why did Robinson crack, for God's sake?'

'Bad scheduling. The day before he'd had to partner Drag in Uneasy Lies the Head that Wears a Crown, and the day before that he had to pretend that his and Joan's marriage was on the rocks.'

'Well, we've simply got to get organized. Get Jones in Planning to work out a complete schedule for everyone concerned. That includes the gang at the Opera Society, the Heart Fund executive, the city councilmen and their staffs, and of course Mrs Bartholomew and the kids as well as everyone here in the administration building and at the plant. Have him lay it out day by day a month at a time, including a likely projection of Drag's appointment schedule a month in advance, with indicated possible deviations. Then have him make a sequential projection of Drag's likely role preferences on the basis of his performances over the past year (I *know* it won't be easy), with six alternative roles for each role in the sequence in decreasing order of probability. He'll have to do the best he can; a year from now we'll have better data and he can make a better projection. Now, when he's done all that have him make up a list of role assignments and give everyone as many as they think they can handle. We'll have to set up a central dispatching office to get people to the right places at the right times, and Drag's secretary and Mrs Drag between them can keep Central Dispatching up to the minute on his location and contacts. OK?'

'I'll see to it right away.'

'If we had had this thing going last night we could have manufactured some excuse for getting Robinson out of there before he cracked. He's no good at Indecisive Executive anyway.'

'But if we keep shuffling people around like that, won't the Old Man get suspicious?'

'We'll try to keep it to a minimum, of course, and when it happens we ought to have a reasonably plausible story (better have Drag's buddy at the *Journal* prepare an index of contingency cover stories), but the most important thing is that as long as we keep feeding him role opportunities he won't pay much attention to what no doubt will, from time to time, develop into a fairly surrealistic sequence of comings and goings.'

The master plan was created and put into operation within the week. For a full year it worked like the well-oiled machine it was.

And then:

'Central Dispatching.'

'Master Plan Control here. Suspend all operations until further notice.'

'What!'

'At fourteen hundred hours this date Bartholomew Drag was admitted to Froehlichkeit Sanatorium for an indefinite period of treatment. Out.'

Three months later

DR HEUSCHRECKE: Make yourself comfortable, Mr Drag. No, not on the couch, please. Take this chair.

DRAG: Heuschrecke, eh? You're the fourth one in three months. But I suppose you know that. I presume you do talk to each other. I almost added 'behind my back.' Ha-ha.

H: Mr Drag –

D: I know, I know. I realize that's a defence you chaps see through in a second, my pretending to make fun of the fact that I'm a paranoid. Quite right, too. So I'll start at the top and give you the whole story, as though you didn't know a thing about my case, just as I did with the others. Right? Right. (*Drag suddenly gets up from his chair, quickly opens a closet door, peers inside, closes the door, and returns to his chair*) I have, you see, the completely irrational belief that I am the object of an elaborate conspiracy. I simply cannot rid myself of the ridiculous notion that everyone I know is humouring me in some way, that there is a concerted effort among my business associates, my employees, my friends, even my family, to keep something from me. I imagine that knowing glances are exchanged, and I find myself interpreting overheard scraps of conversation in such a way as to convince myself that people are planning the most extraordinary things about me. (*Drag empties the waste basket and examines its interior*) And sometimes I fancy that I detect looks of the most extreme exasperation, if not rage, when I seem to catch one of my friends or acquaintances off guard. It is for all the world as though everyone were treating me as a bad-tempered child they were forced to pander to. (*Drag pulls back a corner of the rug and examines the floor beneath it*) But of course these are just my recent symptoms. You want to hear about my childhood. Well, the first thing I remember –

H: Mr Drag, please shut up.

D: What's that? What did you say?

H: Do shut up.

D: What the devil are you saying? You must be out of your mind! By God, I don't believe you're a psychiatrist at all. You don't even know your own, your own –

H: Role, Mr Drag?

D: Well, yes, if you want to put it that way. And believe me I know what your role is supposed to be. I know the drill. I talk and you chaps listen. I think you're a bloody impostor and I demand to see the director at once.

H: 'Drill'? 'Bloody'? You must have picked up that kind of bluster when you were liaison officer with the RAF.

D: What the devil are you talking about?

H: The role you're playing this very minute: Outraged Officer when someone isn't playing the game. You've even assumed something of an English accent. Did you know that?

D: Oh, I get it now. This is some new kind of shock treatment. Well, fine. If you can jolt me out of my paranoia, more power to you.

H: You are not paranoid, Mr Drag.

D: Don't be an idiot.

H: You'll have to take my word for it that I am not.

(*Dr Heuschrecke, who had earlier interviewed the principal officers of the Master Plan, then told Drag the true facts of the case. Drag responded at first by ably performing the role of a man incapable of articulate utterance. Then he gained sufficient poise to speak*)

D: I believe you, Heuschrecke. It's monstrous. I shouldn't be allowed out alone.

H: It must be admitted that you have been something of a trial to your friends and associates.

D: Something of a trial indeed! A bore and a drag; that's my name and that's my game. But can you cure me, that's the important thing? Or is the only way to protect society from me to clap me into a madhouse? What *is* wrong with me, anyway? Is it some new kind of mental illness?

H: Mr Drag, you are not suffering from a mental illness of any kind.

D: Then what on earth is wrong with me?

H: You are suffering from a logical fallacy.

Here the dossier ended and Sneak, smiling broadly, put it down. The next day he was again ushered into Dr Heuschrecke's office.

H: Come in, Sneak, come in.

S: (*handing Heuschrecke the manila folder*) An interesting case.

H: I thought you would find it so.

S: And were you able to effect a cure?

H: If cure can be separated from rehabilitation, then I would say that he was cured but not yet completely rehabilitated, though the prognosis is good.

S: Of course, you wanted me to read this to persuade me that one can perform proprietary as well as assumed roles in games of make-believe.

H: In part, yes. And are you persuaded?

S: Yes, I think so. In fact, Drag's symptoms seem to be a kind of mirror image of my own.

H: Why do you say that?

S: Well, perhaps most strikingly, whereas I was engaged, at the outset, in the deception of other people, in Drag's case other people were engaged in the deception of him.

H: Quite so. And this points to the basic similarity and also to the basic difference between your case and his. The similarity is that both of you needed *situations* in which to perform your roles, that is, the performance of other roles responsive to your own. But each of you achieved this enabling condition in opposite ways. You, at least at first, insinuated yourself into already existing situations by adopting, through imposture, one of the roles which went to make up such a situation. If the others believed you to be the Queen of England, then they would respond in ways appropriate to the Queen and thus enable you to continue your performance of Elizabethan roles. Drag, on the other hand, found himself with roles to perform but a scarcity of situations in which to perform them. He therefore performed his roles even when the situation was inappropriate to their performance. You, on the other hand, performed roles when the role was inappropriate to your own identity. You were an impostor and Drag was a bore, even though your goals were the same; that is, to be performing roles.

S: Yes, that describes it. But how do you account for these quite different approaches?

H: Why, by differences in your backgrounds and, accordingly, in your characters, of course. Drag, after all, was a dedicated boy scout. Honesty was his watchword; deceit was anathema to him. But you were heir precisely to a tradition of professional duplicity. Consequently, when each of you began to achieve that autonomy in your role-performances that we have called playing a game, you adopted quite different strategies to accomplish this autonomous condition. Your performances, Sneak, were like stage performances in the respect that you were trying to stimulate responses from an audience, even though an unwitting audience. You could hardly afford, therefore, to bore them. But Drag did not adapt his roles to his audience; on the contrary, he required them to adapt their responses to his roles. And that, of course, explains why people would rather go to the movies than to church, and why charlatans are more entertaining than honest men.

But these considerations, while interesting, are taking us away from the issues that primarily concern us. We are interested not in the different ways in which you and Drag accomplished your purposes, but in the similarity, indeed the identity, of those purposes. For you played assumed roles and Drag played proprietary roles only because each of you thought that role-performance had to exploit real-life situations, and thus the real-life temperaments of each of you dictated the kinds of role that you would play. But since make-believe can be a game in which the performance of enabling roles is itself part of the game, the distinction between assumed and proprietary roles is irrelevant. Drag is no more constrained by temperament to be 'sincere' in his roles than you are constrained to be 'insincere,' because in a game of pure make-believe the terms have no force.

That distinction is replaced by the distinction between a good and a bad move; that is, between a performance which evokes a response and one which does not. And depending upon the game being played, or upon the state of the game at any given moment, a role might or might not be true to the character of the person who performs it. But what of it? A game is successful just to the extent that it continues to produce responses, not to the extent that it is sincere or insincere. Both of you are therefore in a position to live down your names. You can play games of this kind without being a sneak, and Bartholomew can play them without being a drag.

s: I am convinced by what you say, Dr Heuschrecke, but is Drag, I wonder? His personality strikes me as being altogether more rigid than mine.

H: I really don't think that has much to do with the basic facts of the case, Sneak, although you are quite right, of course, in what you say. For it is characteristic of games that quite divergent personality types can engage in the same game. The fact that so-and-so is a belligerent bastard no doubt differentially colours the game of hockey in which he plays, but this is much less important than the fact that he is a belligerent *hockey* player. But as far as Drag is concerned you can judge for yourself. He is waiting in the ante-room now, I believe. (*Heuschrecke goes to the door and opens it*) Come in, Drag, come in. Mr Drag, I'd like you to meet Mr Sneak.

DRAG: Glad to meet you, Sneak. Heuschrecke has told me something about your case.

H: I trust you don't mind, Sneak?

s: Hardly, doctor, since your bringing us together, I surmise, is in aid of our rehabilitation.

H: Quite right. And the prognosis, gentlemen, is good.

S: It is, is it? (*He produces a revolver*) Don't make a move, Heuschrecke, sit right where you are with your hands on the desk. Drag, I'm not afraid to use this! I want you to get up – not you, Heuschreche, you stay there – and walk ahead of me out to the parking lot. There you will get into the driver's seat of the grey Mercedes parked near the hedge. After that I'll tell you what to do.

D: Very well, but first tell me who you really are.

S: I am Porphyryo Sneak, a retired spy.

D: I just wanted to be sure. And I, so we'll know where we are, am really Sanders of the FBI.

S: Of course you are. Now move, Sanders! (*Sneak and Drag exit*)

H: The prognosis is not good, it's excellent! (*He flips a switch on the intercom*) Please send Mr Skepticus in. (*Skepticus enters*)

SKEPTICUS: Good God, Grasshopper, what are you doing masquerading as a psychiatrist?

GRASSHOPPER: I am not a psychiatrist, I am a –

S: Yes, I know – a physician of philosophy. But why the disguise?

G: Not a disguise, Skepticus, a *nom de guerre*. *Heuschrecke* is German for grasshopper.

S: Oh. Even so, what the devil am I doing talking to you? Grasshopper or Grasshopper-as-Heuschrecke, you are still nothing more than a figment of the real Grasshopper's imagination. How can I be part of the tale that you are at this very moment telling me?

G: Ah, well, Skepticus, who can tell what tale any of us may or may not be a part of? Metaphysics is not really my line, and in any case what difference does it make? In the inquiry we are pursuing it does not matter who says what, or under what curious circumstances, but only whether what is said is cogent and relevant to the issue. So let us now return to that issue.

The important thing is that the moves one makes be good rather than bad—that is, moves which keep the game going instead of terminating the play.

newfeld '78

12 Open games

In which the Grasshopper argues that the two preceding tales have laid the ground for the new concept of the 'open game,' which reveals the original definition as broad enough to cover games of make-believe

SKEPTICUS: I note first of all, Grasshopper, that you arranged things so that Sneak and Drag would live happily ever after.

GRASSHOPPER: And why not, Skepticus? It costs us nothing to suppose that they did, and I like a story with a happy ending.

S: Quite so. And now, Grasshopper, perhaps you would like to draw the moral from their two comedies of error.

G: Certainly. It is that while reverse English can be used to invent or devise games (for that is precisely what Sneak and Drag did), reverse English is no part of what a game essentially is. What Heuschrecke pointed out to his patients was that they could play dramatic games without having to exploit real-life situations, was it not?

S: Yes, it was. Their cure consisted precisely in their coming to accept that fact.

G: But exploiting real-life situations – at least in the ways that Sneak and Drag did – is the same as applying to those situations the principle of reverse English. When Sneak duped others so that he could be playing a part, their being duped was not his primary goal but an occasion or pretext for dramatic impersonation. And when Heuschrecke pointed out to him that he could play his games just as well – if not better – without duplicity, he was also pointing out that he could play these games just as well without reverse English. Similarly, of course, with Drag. When Drag realized that he could play his games without putting English on real life, he stopped being a nuisance and a bore, but he did not stop playing games.

S: So that all we have done in our pursuit of reverse English is to start a hare.

G: Not entirely, I think. For first, I suspect that even if reverse English is not very relevant to *games* as such, it may be highly relevant to *play* as such, and perhaps we can consider that possibility further on another occasion. And second, even if we have started a hare with respect to games, that hare has evidently led us to our real prey. For at first it seemed that the performing of *assumed* roles was the essence of the kind of game we were trying to capture. Then, with the case history of Drag, it became apparent that one could play this kind of game equally well by performing proprietary roles. As Heuschrecke pointed out, the important thing in a game of this kind is not that one assumes a character other than one's own,* but that the moves one makes be good rather than bad – that is, moves which keep the game going instead of terminating the play. And I suggest that the principle of *prolongation* rather than the principle of reverse English is what we were really after all along.

S: Yes, Grasshopper, I took in Heuschrecke's point about prolongation when he made it. But I must say I found it then, as I find it now, a strange thing to say about games.

G: Why is that?

S: Why, because it seems to mark such a striking contrast to the ways in which games are actually played. To work to prolong a baseball game would be to violate the spirit of the game; for example, intentionally to fumble a fly so that the side at bat would not be retired and the game could continue longer. One can do, and no doubt someone or other has done, just that kind of thing, but there is surely something perverse about it.

---

* I realize that this is a somewhat heterodox view of make-believe games. Roger Caillois, for example, in his *Man, Play, and Games* (The Free Press 1961) classes such games as being essentially instances of *mimicry*, one of four basic categories of game that he distinguishes. My view is that while many games undoubtedly *contain* mimicry, and even are appealing *because* they contain mimicry, it cannot be their mimetic component which makes them *games*. Analogously, although athletic games undoubtedly contain bodily actions, it is not that fact that makes them games. For bodily actions are also parts of enterprises which are not games, and so is mimicry. The bodily act of throwing a hand grenade is not usually (and certainly not necessarily) a move in a game, and neither, I submit, is the mimetic act of delivering a line in a play, and yet Caillois seems to regard theatrical performances as examples of mimetic games. I have no quarrel with classifying games in terms of the activities they bring into play (e.g., dramatic moves in contrast to athletic moves); I only claim that such subdivisions can be meaningfully identified only after more basic distinctions have been made, and that the most basic of these is the distinction between enterprises which are games and enterprises which are not.

G: Perverse?

S: Yes, even paradoxical. For anyone who did such a thing would evidently be in the position of prolonging baseball at the expense of genuinely – or at least wholeheartedly – playing baseball. It reminds me of your thesis about games and paradox. Your findings there, it seems to me, are directly relevant to the present issue. For you found that such efforts at prolongation made sense – were not paradoxical – just to the extent that the game or the play in a game was in some way defective, since then such efforts at prolongation could be understood as a kind of piece-meal shoring up of a rickety structure. This suggests, therefore, that Sneak and Drag were at best playing defective games.

G: Not necessarily, Skepticus. In the kind of prolongation which consists in repairing defective games, the efforts to prolong the play are made outside the game, but there may be games whose prolongation is brought about by moves in the game itself. Kierkegaard's Diarist, for example, appears to be playing just such a game, and it is that fact, I suggest, rather than the fact that he appears to be putting reverse English on genuine seduction, which holds the solution to our problem.

S: What do you mean?

G: Well, once he has decided to play the game of Seduction, we find the Diarist cautioning himself against succeeding too soon. The greatest danger to the game is that the girl's ardour for the Diarist may become so great that she will succumb without the necessity for any further campaigning, and so the 'seducer' must, from time to time, throw cold water on her growing passion, though not so much, of course, as to extinguish it altogether. He is, that is to say, continually postponing completion of the game. He keeps moving back the finish line, as it were, so that the race will not end. And when it does end, the Diarist realizes that he will experience not the exaltation of victory but only 'a certain sad satiety.'

S: The Diarist reminds me of the lines in Keats' *Ode On a Grecian Urn*: 'Bold lover, never, never canst thou kiss, though winning near the goal ... / Forever wilt thou love, and she be fair!'

G: Yes, Skepticus, for it expresses the ideal of the Diarist: forever will he chase and she be chased.

S: And chaste.

G: Precisely. For the chase can last just as long as the chastity and no longer. Of course Keats is talking about a realm where the act is safe from consummation because it is frozen in a timeless condition. But the Diarist is really acting, in real time, and so does not have Keats's Platonic option open to him. The best he can do, therefore, is to seek indefinitely to postpone the unwanted denouement, the specious goal. He will inevitably fail, but at least he is doing something about it. And he is doing the best thing he can do, perhaps, if he wants to be acting instead of poetizing, for playing his game may be the best way to realize in time the timeless romantic ideal of Keats. However, I am digressing somewhat from the main point that concerns us.

s: Yes, you are.

G: And that point, Skepticus, is that there appear to be what I should be inclined to call *open games*.

s: Open games?

G: Yes, games which have no inherent goal whose achievement ends the game: crossing a finish line, mating a king, and so on. Games which do have such goals we may call closed games.

s: And the game that Kierkegaard's Diarist was playing was an open game?

G: Yes, except that in playing his open game he was exploiting an already existing goal-governed enterprise – the seduction enterprise – by delaying indefinitely completion of its normal goal. Like Sneak and Drag, the Diarist was playing a two-person, two-role game where the other person was not a player but an unwitting and involuntary performer of the other role. All were exploiting already existing situations (and people) for their own dramatic purposes. But the crucial point is not that they were playing *exploitative* games, but that the games they were playing – which happened to be exploitative – were *open* games, for it is no part of an open game that it must involve such exploitation. This fact became clear in the notorious ping-pong match between Smith and Jones.

s: What ping-pong match?

G: The one I am about to describe to you. Smith and Jones were the two remaining finalists in the celebrated Ming Cup (or Vase) Playoffs, and an enthusiastic group of fans had assembled to watch the match. Smith served, and the first game began. It bade fair to be an excellent match, as the ball flew back and forth between the contestants. But when, after five minutes, no point had been scored, the audience became restless, and some grumbling began to be heard. And after another five minutes it became clear that the players were not trying to score points against one another at all. They were simply trying to keep the ball in play.

'Come on, play the game!' was heard on all sides.

'We are,' Smith called back to the crowd.

'That's not ping-pong,' was the angry rejoinder.

'No, it's not,' put in Jones, 'It's a different game.'

'But how do you decide a winner?' cried another spectator.

'There is no winner in this game,' Smith answered.

'Then how do you tell when the game is over?'

'That's a good question,' was the breathless reply.

Just as the Diarist had made a game out of genuine seduction, Smith and Jones made a new game out of standard ping-pong. And just as the spectators at the Ming Cup Playoffs were nearly beside themselves with frustration at the contestants' failure to get down to the business of scoring points, we may imagine that the object of the Diarist's attentions was affected in a similar way by the dilatoriness of her 'seducer.' But the point is that it is unnecessary to exploit a game of conventional ping-pong in order to do what Smith and Jones were doing. Obviously one can undertake a ping-pong rally simply by deciding forthrightly to do so. One need not pretend – to oneself, to a partner, or to an audience – that one is playing standard ping-pong, just as one may forthrightly play 'Seduction' with a partner who is no more serious about seduction's normal denouement than you are.

s: And you are saying that the games Sneak and Drag were playing are explainable on the model of a ping-pong rally?

G: Precisely. In such a rally A hits the ball to B so that B can hit the ball to A so that A can hit the ball to B, and so on. And just as there was no reason for me to end that sentence (other than its tedium), there is no inherent reason for ending the game that sentence described.

s: But in the games that Sneak and Drag were playing, what corresponds to keeping the ball in play?

G: Keeping the dramatic action going, Skepticus. A delivers a line to B so that B can deliver a line to A, and so on. As in a ping-pong rally, each move is both a response to the immediately preceding move and a stimulus for, or an evocation of, the immediately following move.

s: Except, of course, for two moves, the first and the last. The first move is solely evocative and the last is solely responsive. Otherwise the game could neither begin nor end. And so there could be a game with just two moves:

FIRST MOVE: Never darken my door again!

SECOND MOVE: Very well. Goodbye for ever.

And this would evidently be the shortest game of this kind that one could play, for all other games would have a middle as well as a beginning and an end, and the middle moves would be combination responsive-evocative moves.

G: Quite right. But we should also notice that the shortest open game is also the worst kind of open game that could be played. Giving a response that provides for no further response is just like giving your partner in a ping-pong rally an unreturnable serve. This is admirable in the closed game of ping-pong, but in Ping-Pong Rally it defeats the purpose of the game.

S: Yes, I see that. Now let me see if I fully understand the main thesis that you are advancing. You seem to be saying that there is what might be called the class *open game*, and that games of make-believe are a sub-species of this class.

G: Precisely, Skepticus. I would define an open game generically as a system of reciprocally enabling moves whose purpose is the continued operation of the system. Then, as you suggest, various species can be found within this larger class. Open athletic games, perhaps, would make up one such species, since all of the moves in such games would be bodily manœuvres. Games of make-believe, then, would make up another species, for in them all of the moves would be dramatic performances. Heuschrecke thus correcctly specified a game of make-believe as being 'a reciprocating system of role-performance maximization.'

S: Very well, Grasshopper, I am convinced. Let us, then, begin anew the search for a definition which will cover both open and closed games.

G: That will not be necessary, Skepticus. It is quite clear to me now that the original definition will do very well for both types of game.

S: But Grasshopper, how can that be? The original definition requires that players of games be seeking to bring about a specific state of affairs, but for players of open games, as we have seen, there is no state of affairs they are striving to achieve. They are simply committed to striving indefinitely.

G: I think you say that, Skepticus, only because you are taking an unnecessarily narrow view of what constitutes a state of affairs. In a ping-pong rally there is a perfectly clear state of affairs that the players are striving to achieve. It is the state of affairs which consists in the ball's being in play.

S: Oh.

G: Yes, the ball's being in play is undeniably a state of affairs. And thus our earlier notion that baseball differs from Cops and Robbers as goal-governed activities differ from role-governed activities was incorrect. For it is perfectly clear that games which involve roles can be governed by goals. Seeking to keep a dramatic episode going is to be engaged in goal-governed role-performance. We ought, therefore, to make explicit a rather important point which has been implicit throughout our discussion of open games. It is that the distinction between closed games and open games cuts across the distinction between games like baseball and games like Cops and Robbers. Thus, both a ping-pong rally and Cops and Robbers are open games, even though one involves the performance of dramatic roles and the other does not, and baseball and Charades are both closed games, even though, again, one involves the performance of dramatic roles and the other does not.

s: Very well, Grasshopper, I will grant that open games have goals. Still, that fact by itself does not prove that open games conform to our original definition, for we must also be able to show that such games involve the use of inefficient means in achieving their goals. And I must say that in the typical games of make-believe that we have been considering – as well as in the games that Sneak and Drag were playing – I do not see where the principle of inefficiency comes in.

G: Let us go back to open ping-pong for a moment. We agreed that the goal of the players is to keep the ball in play, did we not?

s: Yes.

G: And they do this by wielding ping-pong paddles, so that their success in achieving their purpose depends upon the skill with which they make their strokes. The slightest mistake in judgment or in execution will result in the defeat of their purpose.

s: That is so.

G: Then isn't it obvious that a more efficient way to keep the ball in play could be devised?

s: You mean that a machine, or perhaps a pair of machines expertly devised for the purpose, could keep the ball in motion much longer and with far less risk of failure than could two humans equipped with ping-pong paddles.

G: Of course. This case is just like the case of a golfer using homing devices on his golf balls in order to achieve a score of eighteen.

s: Very well, but what about games of make-believe? I suppose you are going to drag out one of your machines again, a *Deus qua machina* to resolve the issue.

G: Yes, Skepticus, if a script is a kind of machine. For like the ping-pong machine, use of a script by players of make-believe games would be a more efficient – less risky – means for keeping a dramatic action going than is the invention of dramatic responses on the spot, which is what the game requires. And that is precisely why Sneak rejected Heuschrecke's suggestion that he rehabilitate himself by taking up a dramatic career. From Sneak's radically lusory point of view, acting from a script would be exactly like playing a game of solitaire with a stacked deck.

S: Very well, Grasshopper, I am convinced that our original definition is adequate to account for open as well as for closed games.

G: Splendid, Skepticus. But before we leave this topic, let us return for a moment to some earlier doubts we had about games like Cowboys and Indians and see whether our new understanding of open games can resolve them. We noted that the games of make-believe played by children are characterized by a good deal of argument about the legitimacy of the moves. And I believe our discovery – as I think we may not immodestly call it – of open games provides us with an explanation of that fact. For, I suggest, such disagreement about the moves arises because the players are unclear about the differences between open and closed games. Because cops are 'against' robbers, and cowboys 'against' Indians, children are misled into treating these sets of 'opponents' as they would opposing football or hockey teams, so that the purely dramatic conflict of an open game becomes confused with the genuinely competitive conflict of a closed game. Arguments over a disputed move, therefore, are both muddled and, often, irreconcilable, since one party to the dispute may be tacitly appealing to a rule of an open game while the other party is tacitly appealing to a rule of a closed game. Perhaps it is for this reason that children soon abandon such pastimes in favour of standard closed games.

S: Yes, for that reason and also, I should think, because standard closed games are usually competitive games, whereas open games appear to be essentially co-operative enterprises, and children love to be competing with one another.

G: At any rate the children in our society do. And this prompts me, Skepticus, to hazard the anthropological observation that if societies which place a high value on success through domination are more inclined to emphasize closed games, we might expect societies which place a high value on success through co-operation to be more inclined to emphasize open games.

s: That's an interesting thought, Grasshopper. I wonder if you got the idea from some philosopher or sociologist in the Soviet Union, for I understand that the Russians have interested themselves lately in the study of sport and games. And one might suppose that open games would be seen by them as essentially socialistic, in contrast to the competitive games so popular in capitalistic societies.

G: Your supposition is quite plausible, Skepticus, but quite wrong. There is certainly no distinctively 'socialistic' sport in the Soviet Union or, as far as I know, anywhere else. In Russia hockey is now the national craze. Nor is there any sign that Marxists or any other socialist writers have the least interest in, or indeed awareness of, open games and their possible relevance to an ideological commitment to social co-operation. Of course, Marxists are temperamentally antagonistic to any kind of definitional inquiry, for they look upon definitions as empty abstractions; that is, as things not readily exploitable for doctrinaire purposes. They thus tend to disdain any theory devoid of polemical or ideological promise, which is the reason why some of them preferred, for a time, Lysenkoan to sound genetic theory.

s: And you are claiming that socialists ought to be philosophically or ideologically committed to open games?

G: Perhaps that would be too strong a claim, Skepticus. But one may suggest to those who are interested, or who profess to be interested, in the social determinants of sport and in sports as indicators of social values, that the distinction between closed games and open games might be relevant to those interests. But these speculations, while intriguing – and deserving of further consideration on another occasion – are somewhat tangential to our main concern, which is to test our definition of games. And I take it we agree, Skepticus, that the fact of make-believe pastimes does not pose a threat to that definition. Then let me ask whether you have doubts about the definition's adequacy on any other score?

s: Yes, Grasshopper, I have one. It is a doubt about your account of lusory attitude.

The attitude demanded by radical instrumentali

XIII

, is inevitably one of radical ambivalence.

newfeld '78

13 Amateurs, professionals, and *Games People Play*

In which it is argued that what it is to be a game is independent of the motives anyone may have for playing it, and the 'games' in *Games People Play* are presented as examples of what games are *not*

Your account of lusory attitude [I continued] is expressed by that part of your definition which states that a game player accepts the limitation of means which the rules demand 'just because such acceptance makes possible such activity.' Now consider, if you will, a professional athlete. He is playing hockey, let us say, as a means of earning his livelihood. That is, his reason for playing hockey is to make money. Now, in order to play the game of hockey he must accept the rules of hockey. It therefore follows that one, anyway, of his reasons for accepting the rules is that such acceptance is a necessary condition for earning his salary. But if that is so, then it is false to say that he accepts the rules of hockey 'just because such acceptance makes possible such activity.' And so the existence of professional game players appears to falsify your account of lusory attitude. Unless, of course, you want to maintain that professionals, precisely by virtue of the fact that they do *not* have lusory attitude, are not really playing games. And perhaps that is what you do want to maintain. Perhaps you want to say that when you and I play hockey we are playing a game, but that when Bobby Orr and Ken Dryden play hockey they are working. So my present response to your definition is more of a query than a criticism. Since your account of lusory attitude appears to force you to make a choice between two alternatives – either the admission that your definition is incorrect or else the claim that professional athletes are not playing games – my question is which alternative you wish to choose.

I reject both alternatives, Skepticus [replied the Grasshopper].

Let me first argue in support of the proposition that professionals are genuine players of games and then return to a defence of my formulation of lusory attitude.

Professionals

I would like to make a distinction between what may be called an amateur and what, with some latitude, may be called a professional player of games. By amateurs I mean those for whom playing the game is an end in itself, and by professionals I mean those who have in view some further purpose which is achievable by playing the game. Professional players of chess or bridge as well as professional athletes are obvious examples of such players, but let us extend the term to include players who play games for the sake of any further purpose whatever; for example, to decide an issue ('Let's play a hand of poker to see who goes into town for more beer'), to achieve the greatest good for the greatest number ('You know how I hate bridge, but since you need a fourth I'll play this once'), to gain approval ('Percy joined the football team because Gwendolyn fancies football players'), and so on.

Now, what plausibility there is in the contention that professionals are not really playing games arises, I suspect, from the undeniable fact that the attitudes of amateurs differ from the attitudes of professionals towards the games they play. Thus, although the beer drinkers, the fourth at bridge, and Percy all find the playing of games acceptable undertakings in aid of accomplishing further purposes, it is clear that these purposes are – or at least could very well be – more important to them than the games themselves. If the beer drinkers were to discover an overlooked case of beer, they might have no wish to play the hand of poker; if an enthusiastic bridge player turned up, the unwilling fourth would gladly withdraw; and if Gwendolyn's amatory preferences were to switch from athletic to sedentary objects, Percy might very well resign from the football team. The attitude of the amateur differs from these attitudes because he is motivated by a love of the game rather than by a love of beer, of the general welfare, or of Gwendolyn.

But although the attitudes of amateurs and professionals are markedly different, it is still the case that these differing attitudes are attitudes towards *games*, and not towards something else. In a similar way, Smith and Jones have very different attitudes towards the force of gravity. Smith, who is trying to get a rocket into space, deplores it; Jones, who is trying to return a rocket to earth, applauds it. But these contrary attitudes do not change what it is to be the force of gravity.

It is true, of course, that some things do change with a change of attitude. If playing – rather than playing games – is activity which is always and only undertaken for its own sake, then 'professional player' is a contradiction in terms. On such a view we would be obliged to say that a professional athlete was not *playing*, but we would not be obliged to deny that he was playing a *game*. In the same way, while we would not want to say that a concert violinist was at play during his recitals, we would presumably want to grant that he was playing the violin.

Lusory attitude

But you are wondering, Skepticus, how I can square my belief that professionals are playing games with my account of lusory attitude. For you believe that that account implies that only amateurs can play games, since it holds that anyone who plays a game accepts the rules of the game *just because* such acceptance makes possible such activity. That is, you interpret the phrase 'just because' as necessarily excluding everything except the reason that such acceptance makes possible such activity. And I admit that that is a reasonable interpretation of the phrase 'just because.' But there is another interpretation which is equally reasonable, and so I welcome the opportunity to clarify this part of my definition.

Where A is some action and R is a reason for performing A, you, Skepticus, interpret the phrase 'A just because R' to mean: 1/ R is always a reason for doing A, and there *can* be no other reason for doing A. But I interpret the phrase 'A just because R' to mean: 2/ R is always a reason for doing A, and there *need* be no other reason for doing A. Thus, a player's acceptance of rules because 'such acceptance makes possible such activity' is the only reason he *must* have in playing a game, but it is not the only reason he *may* have. But even the additional reasons he *may* have are limited to a very narrow class. For, as will become evident a bit later on, he can have no reason for accepting the rules which is not also a reason for playing the game. My account of lusory attitude accordingly permits such an attitude to associate, as it were, with other reasons a player may have for playing a game – and therefore for accepting the rules of the game – without that attitude somehow being destroyed or contaminated by such an association. That is, I am not committed to the position that playing a game for some further purpose somehow falsifies the proposition that a game is really being played. Nor, although extra-lusory purposes can be accomplished by playing games, is it necessary either to have or to accomplish such purposes in order to be playing a game; that is, such purposes are no part of the definition of game playing.

My account of lusory attitude is intended to rule out not 'professional' players of games, but the following kind of quasi-game player. Smith arrives at the starting line of the 200 metre finals just as the race is about to begin. He has only that moment learned that a time bomb has been planted in the grandstand at the finish line (which is located on the other side of the oval track at a point directly opposite the starting line), and that it will go off in a matter of seconds. The information has so shocked Smith that he is temporarily bereft of speech and so cannot warn anyone of the impending catastrophe. His first impulse is to run straight across the infield and defuse the bomb, but he sees with dismay that the infield has been fenced off with a high chain-link barrier, evidently to protect spectators and participants from the fifty or so man-eating tigers that roam hungrily inside the enclosure. At the instant Smith realizes that his only hope of getting to the bomb in time is to make a half circuit of the track, the starting gun is fired, and Smith and the other entrants are off and running hard.

Now, I put it to you, Skepticus, that the other runners are playing a game but that Smith is not, and that this is so because the other runners have lusory attitude and Smith does not. Let me explain. Two rules relevant to lusory attitude are at issue in this episode: the rule which requires entrants to begin running at the same time from the same point, and the rule which requires that they do not cut across any part of the infield. Now, through a series of uncanny coincidences, Smith finds himself observing both of these rules. But his reason for doing so is quite different from the reason that the other contestants have for observing the rules. If Smith had arrived at the starting line earlier he would have begun running earlier, and if the infield had not been barred by a tiger-filled enclosure he would have cut straight across the infield. But the other runners, who could have started running before the starting gun was fired, did not do so, and if the infield had been neither fenced nor tiger-infested, they still would have remained on the track. That is, they accepted the rules just because they wanted to participate in a competitive game. But Smith acted within the constraints because that was the only way he could get speedily to the bomb. Clearly his attitude towards the rules was not that they made possible a foot race, for if he had found his voice or if the infield had been safe and clear, he would not have been running around the track at all.

Smith's attitude, I suggest, puts the difference between amateurs and professionals into proper perspective. For although professionals and amateurs admittedly have different attitudes towards the *games* they play, they have the same attitude towards the *rules* of those games, an attitude which is the opposite of Smith's. For let us suppose that the other runners had all been professionals rather than amateurs. They still, unlike Smith, could not jump the gun or cut across the infield without utterly defeating their professional purposes, for it is excellence in playing a game, and in playing a game alone, which serves those purposes. They are *using* a game, to be sure, but they are using a game by playing it. Smith is using a game without playing it. They are contestants: he is an opportunist. And so when Smith, after getting to the finish line ahead of the other runners and defusing the bomb, is disqualified from the race for having interfered with another runner at the second turn, he simply chuckles to himself and goes about his business. The same sort of attitude is illustrated somewhat more poignantly in a cartoon which appeared in, I believe, *Playboy* magazine. It shows a flock of maidens being cast into a fiery pit before some pagan altar, while a number of others are waiting their turn in line. One of these turns to her neighbour and remarks, 'The joke's on them. I'm not a virgin.'

I believe I have satisfactorily defended the definition against the dilemma you advanced against it, Skepticus, but before we leave the question of lusory attitude, I would like to make a bit more clear, if I can, what it means to play a game as an instrumentality, that is, to be what we have called a game-playing professional. For there is some danger that the conclusion we have drawn that games can function as instruments without thereby ceasing to be games may become confused with the proposition that games are *essentially* instruments of one kind or another. Our view of games occupies a middle position between two extreme positions which we reject: what may be called, on the one hand, *radical autotelism* and, on the other hand, *radical instrumentalism*. Radical autotelism is the view that unless games are played solely as ends in themselves, they are not really games, that is, that amateurs alone are playing games. We have already rejected radical autotelism in arguing that professionals, too, are genuinely playing games. Radical instrumentalism is the view that games are essentially instruments, and we also reject that view because, to begin with, radical instrumentalism would evidently hold that Smith was playing a game. But since one widely read authority on games seems to be a radical instrumentalist, perhaps we should take a closer look at that doctrine.

Radical instrumentalism

What does it mean for a game to be essentially an instrument for some further purpose? It means that in the absence of such a purpose nothing worth-while – or, indeed, intelligible – can be going on. Games so conceived are, of course, quite different from any of the cases of professional game playing that we have considered. For in those cases, although games were used for further purposes, those games were, and were known by their players to be, different from and, so to speak, detachable from, the purposes to which they were put. Thus, any one of those games could be put to quite different purposes, and a number of different games could be put to the same purpose. The view I am calling radical instrumentalism is, in effect, the denial that games have such detachability and versatility. Games are in this view conceived as having their instrumental goals built into them or, in the language of our definition, games are viewed as being essentially instruments for the achievement of prelusory goals.

But such a view of games appears to be self-defeating, for excessive dedication to the attainment of prelusory goals has the effect of destroying the games in which those goals figure. Thus Smith was not playing a game for the same reason that cheats are not playing games. Both are pursuing a goal whose attainment overrides obedience to the rules. The only difference between them is that the cheat actually breaks the rules of the game, while Smith, although he does not break any rules, would if he could. And Ivan and Abdul failed to create a rule-less game precisely because a 'rule-less game' is an activity in which achievement of the prelusory goal has become the overriding concern of the participants, and thus fails to be a game.

The queerness of radical instrumentalism becomes even more evident if we turn from the unorthodox behaviour of cheats, Smiths, Ivans, and Abduls and consider conventional games from the viewpoint of that doctrine. Chess becomes essentially a procedure for acquiring chessmen, hockey essentially a procedure for getting rubber disks into nets, and foot racing essentially a procedure for breasting tapes. The queerness of the doctrine lies in the fact that if games are essentially procedures of this kind, then they are as unsuited to their purposes as they could possibly be. And an obvious corollary is that one of the worst ways to achieve some practical objective – building a house, closing a business deal, gaining sympathetic attention – would be to make that objective the prelusory goal of a game.

The attitude demanded by radical instrumentalism is inevitably one of radical ambivalence. It is perhaps the 'odd volitional posture' Kolnai erroneously attributed to genuine players of games. This can be seen by making some modifications in the bomb-defusing episode. Let us replay that race, but with the tiger-filled enclosure eliminated and under the supposition that Smith is keen upon winning the race as well as upon getting to the bomb in time. As the starting gun is fired, he believes that these two purposes can be accomplished by the same means, that is, by running as fast as he can around the track. But scarcely has he left the starting blocks when he realizes that these two goals are in conflict with one another. For he sees that cutting across the infield is a better way speedily to defuse the bomb than is running all around the track. Faced by this choice Smith will do one of three things. 1/ If he values winning the race more than defusing the bomb, he will stay on the track. 2/ If he values defusing the bomb more than winning the race, he will cut across the infield. 3/ If he values each of these things equally, he will be reduced to a state of gibbering indecision.

Radical instrumentalism, therefore, is a theory of games which needs only to be understood in order to be shunned, for it cannot be put into practice. Because of the equal but irreconcilable demands of the game and of what may be called life, although it is possible to meet the demands of the game or of life or of neither, it is not possible to meet the demands of both.

'Games People Play'

If the games played in Eric Berne's *Games People Play* * are really games, then Berne is an exponent of this incoherent theory. For the players of Bernean games are playing them only in order to gain what Berne calls 'strokes,' a stroke being a unit, so to speak, of social recognition. It is true that the attitude towards games that I have called professionalism also permits the playing of games in order to gain recognition; indeed, the best athletes are probably motivated by this consideration most of the time. But while an athlete gains recognition as the *result* of performing some feat, for Berne's players of games the feat performed *is* the gaining of recognition. Or in the language of my theory, the gaining of recogni-

---

* Eric Berne *Games People Play* (Grove Press 1967); further reference to this work will be made in the body of the text.

tion is the prelusory goal of the games that Berne's people play.

Although that difference is the crucial difference between Bernean games and the things conventionally called games, two other important differences figure in his account of games. Let us note them now for future reference: the first is that Berne's players play games only because they are more or less neurotic, and the second is that the games they play are more or less unconscious. Since it is my contention that radical instrumentalism is a self-contradictory principle, it will be interesting to see what happens to Berne as he applies that principle to the interpretation of the behaviour of his subjects. For we may confidently predict that, as with Smith in the replayed race, one of three things will happen: 1/ it will become evident that because his subjects are playing games, their behaviour is dysfunctional for satisfying their neurotic needs for recognition, or 2/ it will become evident that because their behaviour does satisfy those needs, they are not really playing games, or 3/ it will become evident that Berne is talking gibberish.

We may take what Berne calls 'Schlemiel' as a representative example of a Bernean game. Here is how Berne describes it:

The moves in a typical game of 'Schlemiel' are as follows:
    1W   White spills a highball on the hostess's dressing gown.
    1B   Black (the host) responds initially with rage, but he senses (often only vaguely) that if he shows it, White wins. Black therefore pulls himself together, and this gives him the illusion that he wins.
    2W   White says: 'I'm sorry.'
    2B   Black mutters or cries forgiveness, strengthening the illusion that he wins.
    3W   White then proceeds to inflict other damage on Black's property. He breaks things, spills things and makes messes of various kinds. After the cigarette burn in the tablecloth, the chair leg through the lace curtain and the gravy on the rug, White's Child [Berne means by this the child in all of us] is exhilarated because he has enjoyed himself in carrying out these procedures, for all of which he has been forgiven, while Black has made a gratifying display of suffering self-control. Thus both of them profit from an unfortunate situation, and Black is not necessarily anxious to terminate the friendship.

As in most games, White, who makes the first move, wins either way. If Black shows his anger, White can feel justified in returning the resentment. If Black restrains himself, White can go on enjoying his opportunities. The real payoff in this game, however, is not the pleasure of destructiveness, which is merely an added bonus for White, but the fact that he obtains forgiveness. (p. 114)

'Schlemiel' admittedly bears some resemblance to genuine games. There are moves and counter-moves and there is what Berne calls a 'payoff.' But I think, Skepticus, that the similarity pretty much ends there. Notice, for example, Berne's very odd remark that 'as in most games, White, who makes the first move, wins either way.' That is an odd thing to say about games, because such a state of affairs is ordinarily, if not invariably, the mark of a seriously defective game, as we had reason to observe on another occasion. Such a game would be a Parker Brothers reject. Or if football were such a game, then the team that won the toss would be assured of victory, so that football could be replaced by coin flipping.

We may also notice how Berne is using the word 'payoff.' Now as far as games are concerned 'payoff' is ambiguous. The joy of victory – or even the satisfaction of playing a losing but superb game – might both be considered payoffs. Or some additional reward for winning might be considered a payoff. The ambiguity is revealed in an exchange depicted in a *Punch* cartoon between the father of a marriageable daughter and a young man.

FATHER: Whoever marries my daughter gets a prize.
YOUNG MAN: Jolly good. Will it be a cash award or just a trophy?

The payoff in Berne's games, it is clear, is like a cash award or a trophy and unlike the satisfaction of a well-played game, as I believe is convincingly established by the following exchange we may imagine as occurring between Sam Schlemiel, an avid player of 'Schlemiel,' and a friend.

FRIEND: Why don't you play 'Schlemiel' with Abe Adult rather than with Suzy Schlemazl? (The 'Schlemazl,' according to Berne, is the natural victim of the Schlemiel.)
SAM: Why should I? Suzy is perfect for my purposes. She always forgives me at once, no matter how outrageous or how numerous my transgressions.
F: Precisely. It's just like shooting fish in a barrel. But you'll get some really good play with Abe. He's not all that easy to fool.
S: Good heavens, that's not a reason for playing with him. Do you suppose I play 'Schlemiel' for the fun of it?
F: I certainly thought you did. Aren't you playing a game?
S: Not *that* kind of game. I'm not out for *sport*, old man. Sport I can do without. But if I don't get my strokes I'll go to pieces.

A player of 'Schlemiel,' it is clear, values the strokes, not the activity directed to producing the strokes. For if either of two conditions existed, the player of 'Schlemiel' would not play it: 1/ if he was getting sufficient doses of forgiveness in the ordinary course of events, or 2/ if he had overcome his neurotic need for strokes of this kind.

But such an attitude is utterly unlike the attitude of golfers and chess players. For suppose, carried away by Berne's thesis, we uncritically accepted his contention that games are unconscious devices for the satisfaction of neurotic needs. We have a friend who is devoted to golf, and so we try to cure him of his mania. We set him to filling holes in the ground with golf balls in his back yard. After a week of this we call upon him and confidently inquire whether he has now been cured of golf. With a pitying grimace he flings his golf clubs into the trunk of his car and speeds off to the country club. Or we try to cure another friend of chess by inundating him with chessmen. They arrive at his house by mail, by special messenger, by van. When we call upon him later to observe how his convalescence is progressing, we find that he has moved his chess table out to the front porch because his house is so full of chessmen that there is no room to play chess.

If the games that Berne's people play are really games, then Berne is committed to absurdities of this kind, and he is revaled as talking gibberish. But Berne is not really that crazy, and his subjects, although admittedly not paragons of mental health, are not that crazy either. For of course the things that Berne calls games are not games at all. Indeed, no one could be less interested in games like chess and golf than Berne. Even though he borrows 'White' and 'Black' from chess, neither that game nor anything like that game is the model which guides his analysis of social behaviour. His model is not any kind of game, but the confidence 'game' – that is, a certain kind of trickery and deceit, as Berne makes quite clear:

A game is an ongoing series of complementary ulterior transactions progressing to a well-defined, predictable outcome. Descriptively it is a recurring set of transactions, often repetitious, superficially plausible, with a concealed motivation; or, more colloquially, a series of moves with a snare, or 'gimmick.'

And so Berne concludes, not very surprisingly, that 'every game ... is basically dishonest' (p. 48).

It is true, of course, that trickery and deceit are part of many games. The feint in fencing and boxing, misdirection in chess and in various card games, the 'deke' in hockey, the curve in baseball – all are efforts to mislead in order to gain an advantage. But it is not these manœuvres that make the activities in which they occur games; it is the constitutive rules of those games which make these kinds of misdirection the useful manœuvres that they are. But to call any deceptive move whatever a move in a game is to court, if not to become wedded to, quite unnecessary confusion.

You may be thinking, Skepticus, that I am making a great deal of fuss over what is nothing more than a verbal quibble. What's in a name, after all? Three considerations have prompted my interest in Berne's use of the word 'game,' and while two of these considerations lead me to deplore that usage, the third is a consideration of a different kind.

1/ There is no point in being muddled when it is just as easy to be clear. I suggest that the phrase 'working a con' could be substituted for the phrase 'playing a game' wherever the latter occurs in Berne's discussion with no loss in meaning and with a definite gain in clarity.

2/ Linguistic muddles can have practical consequences. Thus, calling war a game, as Berne does (p. 50), is not only jejune but also, perhaps, dangerously misleading. For it suggests that wars are in principle as easily avoidable as is the pursuit of some popular but destructive sport. Thus, the mayhem of the 1973 Indianapolis Five Hundred was avoided in the 1974 running by the introduction of rules which resulted in a race with zero injuries to drivers and spectators. People who think of wars as games may be misled into supposing that wars are subject to the same kind of reformation, if only it is pointed out to statesmen and generals that some modification of the rules of war is desirable, in view of the fact that under the present rules people are actually getting killed. Or if war were a game, it would be perfectly appropriate to congratulate a general on his sportsmanlike conduct in declining to gain an advantage over his opponent by launching a surprise attack. As Ivan and Abdul would say, 'Tell it to Moshe Dayan.'

3/ But in fact my primary purpose is not to chastise Berne (whose analysis of certain common forms of social behaviour may be first rate) for his casual use of the word 'game,' because he is by no means the only offender; there is a good deal of loose talk about games these days. No, Berne is essentially interesting to me because his thesis, which may be sound social psychology, beautifully exemplifies, if it is regarded as a thesis about games, the incoherence of radical instrumentalism.

A final point of interest to us emerges from Berne's account, Skepticus. At the end of his book he contrasts a condition he calls 'autonomy' with the neurotic dependence characteristic of people who play 'games.' And one of the chief features of this autonomous condition is that 'it means liberation, liberation from the compulsion to play games' (p. 180). This compulsion to play Bernean 'games' is, I suggest, just like the compulsion that ants have to work, except that ants work in order to secure their physical survival while Berne's people play 'games' in order to secure their psychological survival. And just as ants would have no reason to work if they achieved a condition of economic autonomy (i.e., independence), so Berne's players (Berne is saying) would have no reason to play 'games' if they achieved a condition of psychological autonomy. Now this is very interesting to me, Skepticus, for it brings to very sharp focus the irreconcilable difference between the things Berne calls games and the things I call games. For I suspect that playing (genuine) games is precisely what economically *and* psychologically autonomous individuals *would* find themselves doing, and perhaps the *only* things they would find themselves doing.

Those were the Grasshopper's final words to me in defence of his definition of games.

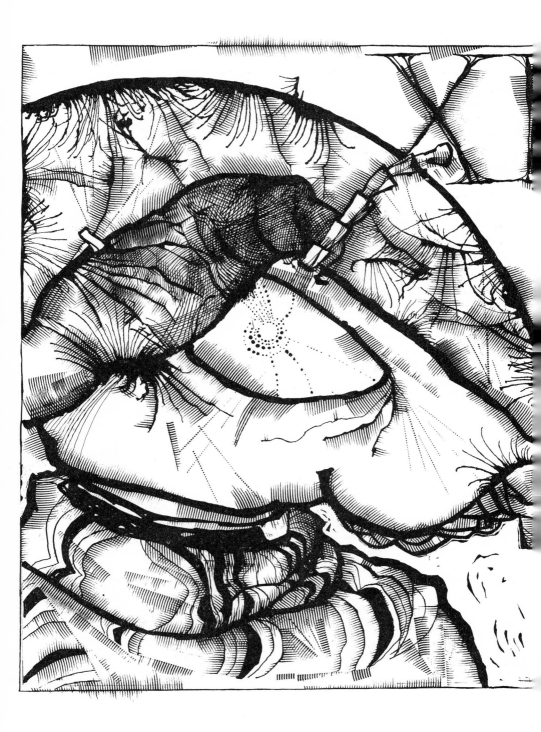

14 Resurrection

In which the flashback ends with the miraculous return of the Grasshopper to resume discussion with his disciples, who have utterly failed to solve the riddles he left them when he died

SKEPTICUS: Prudence, it is now mid-November. We have pondered the Grasshopper's deathbed riddles in the light of his theory of games for more than a fortnight and yet we seem no closer to a solution than ever. So I think we shall have to abandon the game we so eagerly anticipated playing, for evidently we cannot even begin it. We might as well have been trying to play tennis with a two-hundred-pound tennis ball.

PRUDENCE: I'm afraid I must agree with you, Skepticus. I feel as helpless and stupid as Dr Watson after Holmes has given him all the clues and then sits back with a superior smirk at his friend's blank countenance.

S: Quite so. For even knowing what a game is – or at least knowing what the Grasshopper believes a game to be – seems to have no bearing whatever on the Grasshopper's apparent conviction that the life of the Grasshopper must be a life devoted to game playing rather than to trombone playing.

P: Or to intellectual inquiry or to love. For surely these things enjoy as much 'autonomy' as does the playing of games. Why must a life freed from the necessity to work be identical with a life dedicated to games?

S: That is the question precisely. If only the Grasshopper were here there are some objections I could put to him, and then perhaps we could begin to see the direction of his thinking. (*There is a kind of soft scratching at the door*)

S: I'll get it, Prudence. (*He opens the door to find the Grasshopper, wearing an air of some bewilderment, standing on the stoop*) My God! Prudence, it's the Grasshopper! (*Prudence rushes to the door*) Grasshopper, you're alive!

P: It's a miracle!

GRASSHOPPER: Evidently.

P: Come in and sit down. You look quite dazed.

G: Thank you, Prudence. I do feel a bit giddy.

P: But what happened? How do you account for your resurrection?

G: I don't suppose one does account for miracles, does one, since they are unaccountable occurrences. Still, we may take note of the fact that there is in progress an unusually fine Indian summer, so my presence among the living should perhaps be regarded as more a stay of execution than an outright reprieve.

S: But *how* did it happen?

G: I hardly know. I remember bidding you and Skepticus farewell, and then oblivion – until about half an hour ago.

P: And what happened half an hour ago, Grasshopper?

G: Why, of all things, I found myself seated in a grandstand watching a cricket match.

S: A cricket match!

G: Yes. The football was on the fifty-yard line.

S: (*laughing*) Grasshopper, I'm afraid you're a bit addled. Were you watching a football match or a cricket match? Give your head a good shake. It will help to clear your mind.

G: I have no intention of giving my head a good shake. *My* mind is quite clear, thank you. But you were never any good at riddles, I now remember, so let me explain. 1/ The game I was watching was football. 2/ The players on the two teams were crickets. 3/ I repeat that I was watching a cricket match and the football was on the fifty-yard line.

S: So you were resurrected into a pun. That's rather funny.

G: Only faintly so. And such a level of humour, I must say, is just what one might expect of an Agency which found it amusing to perform a resurrection in the first place. A riddle's turning on an obvious pun is just about as witty as the practical joke of resurrecting the dead.

P: And do you really believe, Grasshopper, that there *is* some Agency which controls our destinies?

G: More specifically, I believe that there is some Author who writes our dialogue.

P: Why do you say that?

G: Well, he has given himself away twice already, hasn't he? How on earth could Skepticus here have been admitted to Heuschrecke's consulting room?

s: Yes, that was odd, as I believe I remarked at the time.

G: Odd? It is impossible outside of fantasy fiction. And what about my resurrection? How many people get resurrected these days?

s: But why do you suppose he tipped his hand?

G: The arrogance of power. He can afford to tip his hand because he holds all the cards. And he has, as we have noticed, a fairly primitive sense of humour, so that he thinks it clever to jumble up first, second, and even third order narrative levels as, of course, he is doing right now in having me say the things about him that I am saying.

s: Hold on, Grasshopper, you are giving me an attack of vertigo.

G: Not vertigo, Skepticus, but that other Latin form of dizziness.

s: What is that?

G: Pirandello.

s: Ah, the feeling is passing. What? Oh, yes, Prudence, thank you. Neat, please, with just a little ice. What were you saying, Grasshopper?

G: Just that you were suffering a mild attack of Pirandello, but I see that you have recovered.

s: Yes, I have. Now, Grasshopper, let me see if I understand you. You seem to be saying that this Author you speak of is playing some kind of game with us.

G: If you mean that he is trifling with us, then I agree that he is to some degree doing that. But whether he is playing a game, and with whom, is quite a different question.

p: Do you think he is playing a game, Grasshopper?

G: I think he may be, although it is perfectly possible that he is not.

s: That's highly illuminating.

p: Skepticus! Why do you say that, Grasshopper?

G: We have noticed that our Author sometimes tips his hand, so we must ask ourselves what this hand is that he tips. What, that is to say, is he up to?

p: And what do you think he is up to.

G: I think he is writing a treatise on the philosophy of games.

p: Brilliant, Grasshopper!

G: Elementary, my dear Prudence.

s: Assuming that there really is such an Author, I must admit that your hypothesis does fit all the facts. But how does a game, which he may or may not be playing, come into the picture?

G: Well, if he is writing the kind of treatise I have suggested, why doesn't he simply argue his position in a straightforward manner, like other authors who produce philosophical works? Why introduce the fanciful complication of presenting his thesis through the mouths (or at least from between the mandibles) of three insects? And why add the burden of having to remain in more or less consistent allusive touch with Aesop, Socrates, and *The New Testament*? Why do all that?

s: And you are suggesting that his reason might be that he has created a game by imposing upon his philosophical enterprise a constitutive rule which requires him to express his arguments only within those literary constraints.

G: Precisely. For he certainly has at his disposal simpler and more efficient means for cogent argumentation; most notably, the use of unadorned syllogisms. So his refusal to express himself in a plain expository style is perhaps no different in principle from someone's setting out to write an entire book without using the letter *e*. And if that *is* the kind of thing that our Author is doing, he may even be competing with another Author who is similarly engaged. Or he may be trying to win a bet. Or he may simply be playing the game for the fun of it. On the other hand, of course, he may be doing nothing of the kind. The dramatic and allusive style of his presentation may serve a quite different purpose. He may believe that with that style his work will be likely to secure a larger (and more profitable) readership than it would without such literary embellishments. Or he may believe that even if his arguments do not convince, they may at least entertain. Again, he may (incorrectly) believe that the cogency of his arguments is inseparable from their dramatic and allegorical presentation.

P: And do you believe, Grasshopper, that he is doing one of these latter things, or do you believe he is playing a game?

G: I think there is *some* evidence to suggest that he is *not* playing a game. Earlier I surmised that he mixed up narrative levels from time to time simply because he found it amusing to do so, perhaps out of the sheer exuberance of power. I mean, for example, his representing Skepticus as talking to Heuschrecke. But now I realize that there may be a reason other than his own amusement for the literary liberties our Author is sometimes inclined to take. Such behaviour may be his way of conveying to the reader a message of the following kind: 'Please don't get the idea, dear reader, that I am playing some kind of game which requires me to convey my philosophical ideas always and only within a consistent narrative form. It is true that I *prefer* to do that, for I am trying to write a book which is not too boring to read. But the expression of my argument is of paramount importance to me, and if there should arise, in the writing of it, a conflict between the presentation of that argument and the narrative form in which I have chosen to express it, then it is the form which must give way. For I am entirely ready to disrupt the narrative form at will. Behold.' And he proceeds to prove his point by having Skepticus walk into Heuschrecke's consulting room.

P: So you are satisfied that he is not playing a game?

G: No, Prudence, I am not. For it is equally possible, I should think, that his mixing up of narrative levels is simply a literary lapse on his part. 'Even the great Homer,' as Horace reminds us, 'sometimes nods.' And our Author is no Homer. And therefore, since we cannot decide the issue either way, I suggest that we abandon these theological speculations and return to the matter at hand.

P: I agree. Where were we?

G: Well, let me think. Ah, yes. By resurrecting me into the cricket pun, our Author had got us talking about riddles, no doubt for some philosophical or dramatic purpose of his own.

S: Right. And speaking of riddles, Grasshopper, Prudence and I require your assistance on a much more difficult and important one than the riddle of the football-playing cricket teams.

G: See, I told you so. But what riddle is that, Skepticus.

S: Why, the riddle you bequeathed to us when you died. Surely you remember that?

G: Ah, yes, to be sure. I was telling you about my dream in which everyone alive was an unconscious game player.

P: That's right, Grasshopper. Please tell us at once the meaning of the dream.

G: Gently does it, Prudence, gently does it. I'm not sure I know the meaning of the dream myself, for –

P: Grasshopper, don't say that! I'm dying of curiosity!

G: For, as I was saying, I *have* been dead for several weeks, so perhaps my mind is not, after all, as clear as it might be. Still, if Skepticus will assist me by posing his usual acute questions, that will no doubt help to focus my thoughts and so render my mind once again the finely tuned analytical instrument we know it to be.

S: Splendid, Grasshopper, splendid! Then let me begin by asking you to consider a proposition you earlier advanced as a basic principle of existence; namely, that the life of the Grasshopper – that is, a life devoted to play – is the only justification there can be for work, so that if there were no need for work, we would simply spend all of our time at play.

G: Yes, Skepticus, I recall making that claim.

s: Good. Now consider. Since you use the terms 'work' and 'play' as logical complements of that class of things which we may call 'intentional behaviour,' you evidently conclude that if an activity is not work, then it is play, and *vice versa*. But *prima facie*, at least, that is an unconvincing dichotomy. For example, passing the time of day with a colleague appears to be neither work nor play, and attempting to solve a double crostic appears to be both work and play. As descriptions, therefore, the words 'work' and 'play' seem not to designate sub-sets of intentional behaviour which are either exclusive of one another or exhaustive of the set which includes them.

G: My dear fellow, that's extremely well put. I agree with everything you say.

s: You do?

G: Certainly. My conclusion, however, is not that I gave you poor descriptions of work and play, but that I did not give you descriptions at all. I was using the words 'work' and 'play' stipulatively rather than descriptively. I meant by 'work' activity which is instrumentally valuable, and by 'play' activity which is intrinsically valuable. What play 'really' is, and indeed whether play is definable (other than stipulatively) at all, are questions that need not concern us now. Although, as it happens, I do have definite views on the subject, as I believe I intimated to you, Skepticus, in connection with our consideration of the principle of reverse English. Perhaps, time permitting, we can take up that topic on another occasion.

s: I should like nothing better, Grasshopper.

G: Good. But for now it is clear, I take it, that by 'play' I mean nothing more than all of those activities which are intrinsically valuable to those who engage in them.

s: Yes, and I am delighted to hear you say that, for it clears up one difficulty. Now here is another. I take it that the life of idleness which you exemplify by being the Grasshopper, and which you go about recommending to anyone who will listen to you, is a life devoted exclusively to intrinsically valuable activities.

G: That is so.

s: Then surely, Grasshopper, a life devoted to game playing cannot be identical with that life. For although game playing as you define it *is* an intrinsically valuable activity, it is certainly not the case that all intrinsically valuable activities are games. One may also value for their own sake such things as scratching an itch or listening to a Beethoven quartet, but their being intrinsically valuable does not make such things games.

G: Once more, Skepticus, you are perfectly correct. And your questions have quite cleared my mind, so that I believe I can, with your continued interlocutory assistance, resolve those perplexities which my dying words occasioned. Unless, of course, you and Prudence would rather work out the answers for yourself. That would be to make a kind of game out of the task. (*Skepticus and Prudence exchange despairing glances*) For you would then voluntarily be eschewing superior means for getting the answers just in order to be using your *own* wits toward that end.

S: Death, I see, has not noticeably mellowed your sarcastic nature.

G: Mellowed! I should think not. Dying was the most exasperating thing that has happened to me in my entire life. In any case, what about my proposal? And now that I think of it, we might add just one more limitation to make the game more interesting.

P: What is that?

G: A time limit, of course. What shall it be, then, a day, a week? Skepticus? Prudence? What do you say?

S: Thank you for making the offer, Grasshopper, but I think not. For a time limit, indelicate as it may be to bring it up, is not completely within our power to set. After all, you have died once already, and Indian summers have a habit of ending rather abruptly. Since you may die again at any time, I urge you to begin at once and tell us the solution.

G: As I had reason to observe earlier on, the two of you are not quite grasshoppers yet. A true grasshopper would fairly have leapt at the opportunity to play a timed game where the length of the game is kept from the players.

P: But surely, Grasshopper, not at the risk of losing for ever that knowledge which is alone capable of justifying his existence.

G: Ah, Prudence, but it is part of the thesis that I shall presently expound to you – time permitting, as Skepticus (triumphing over my sensibilities) has pointed out – that a *true* grasshopper would sacrifice anything and everything to be playing games. In fact, however, a true grasshopper would *not* be risking loss of the knowledge to which you allude. A true grasshopper already knows what justifies his existence, for a true grasshopper – and this will mystify you further, Skepticus – already knows everything there is to know. But that is getting a bit ahead of the story.

s: Grasshopper, the floor is yours. Please, for heaven's sake, begin.

p: Yes, Grasshopper, please.

g: Very well, my friends, I shall.

Resolution

newfeld '78

15 Resolution

In which the Grasshopper solves
all of the riddles by outlining
a picture of Utopia

GRASSHOPPER: The solution of the riddle has three chief elements. They are 1/ *play*, as we have stipulatively used that term to designate any intrinsically valuable activity, 2/ *game playing* as I have defined it, and 3/ what I should like to call *the ideal of existence*. By the ideal of existence I mean that thing or those things whose only justification is that they justify everything else; or, as Aristotle put it, those things for the sake of which we do other things, but which are not themselves done for the sake of anything else. Now, I believe that the two of you have assumed I am making the claim (a claim which is, I agree, *prima facie* plausible) that play is identical with the ideal of existence. But the position I shall attempt to establish requires a modification or interpretation of that claim. This position can be expressed by two related contentions. The first is that play is necessary but not sufficient adequately to account for the ideal of existence. The second is that game playing performs a crucial role in delineating that ideal – a role which cannot be performed by any other activity, and without which an account of the ideal is either incomplete or impossible.

In order to support these contentions I would like to use the kind of device Plato used in trying to get at certain characteristics of the human psyche. If we look at the state, said Plato, we will find there the magnified extensions of the characteristics of the psyche that we are seeking, and, being magnified, they will be easier to recognize. Somewhat similarly, I would like to begin by representing the ideal of existence as though it were already instituted as a social reality. We will then be able to talk about a Utopia which embodies that ideal –

that is, a state of affairs where people are engaged only in those activities which they value intrinsically.

Let us imagine, then, that all of the instrumental activities of human beings have been eliminated. All of the things ordinarily called work are now done by wholly automated machines which are activated solely by mental telepathy, so that not even a minimum staff is necessary for the housekeeping chores of society. Furthermore, there are so many goods being produced so abundantly that even the most acquisitive cravings of the Gettys and Onassises of society are instantly satisfied, and anyone who wishes may be a Getty or an Onassis. Economically, the condition of man is a South Sea island paradise, where yachts, diamonds, racing cars, symphonic performances, mansions, and trips around the world are as easily plucked from the environment as breadfruit is in Tahiti. We have, then, eliminated the need for productive labour, for the administration of such labour, and for a system of financing and distributing such production. All of the economic problems of man have been solved for ever. Are there any other problems? There are indeed. There are all of the interpersonal problems which do not depend upon economic scarcity.

Let us, then, further imagine that all possible interpersonal problems have been solved by appropriate methods. Let us suppose that psychoanalysis has made such giant strides that it actually cures people, or that all the various kinds of group treatment have proved successful, or that some quite new development in socio- or psychotherapy or in pharmacology has made it possible to effect one hundred per cent cures for all psychic disturbances. As a result of these developments there is no longer any competition for love, attention, approval or admiration, just as there is no longer any strife in the acquisition of material goods. Perhaps a single example will serve to illustrate the state of affairs in question. Let us take the case of sex. Under present conditions, there is a short supply of willing sexual objects relative to demand. And it may be surmised that the reason for this is the prevalence of inhibitions in the seekers of such objects, in the objects themselves, or in both, so that great expenditures of instrumental effort are required in order to overcome them and thus get at the intrinsic object of desire. But with everyone enjoying superb mental health the necessity for all this hard work is removed and sexual partners are every bit as accessible as yachts and diamonds.

SKEPTICUS: But what about love, approval, attention, and admiration,

Grasshopper? Even if it is not necessary to *compete* for these things in Utopia people would still have to *work* to achieve them.

G: On the contrary, Skepticus, many people seem to believe that the kind of love, attention, and admiration alone worth having is just the kind that one ought *not* to work at.

S: Yes, but many other people, such as marriage counsellors, take a quite different view. They are always saying things like, 'You have to *work* at your marriage, you know.'

G: Yes, but what does this 'working at' mean in the case of marriage or, for that matter, in the case of any other intrinsically valued relationship between people? Does it not mean, essentially, being tolerant of, and helpful with respect to, one another's social and psychological shortcomings? But in Utopia we are supposing that there are no such shortcomings to be tolerant of. Furthermore, whether it is or is not the case that in Utopia one will have to work at something in order to gain love and admiration, it cannot be love and admiration at which one works. We admire a person who works hard, let us say, at teaching. But we admire him because he works hard at teaching, not because he works hard at being admired. I suggest that for convenience we lump together under the word 'approval' all of the pro-attitudes we have been talking about and then ask whether there is anything at all that our Utopians could do to gain approval.

S: Very well. First, then, it is clear that they cannot gain approval by their economic industry, since there is no need for such industry. And I take it that we must also rule out approval for governing well, since with no competing claims for goods requiring legislation and adjudication, there is no need for government. What seems to be left for approval is excellence in moral, artistic, and intellectual accomplishment. Do you agree?

G: For our present purpose, at any rate, I think your list will do. Let us consider moral goodness first. Will you agree with me that moral action is possible only when it is morally desirable to prevent or to rectify some wrong or evil that is about to be or has been done somebody?

S: Yes, I agree with that.

G: But we are also agreed, are we not, that in Utopia no evil or wrong can befall anyone?

S: Yes, that is true of Utopia by definition, since Utopia is just a dramatization of the ideal of existence, and evil and wrong-doing are obviously inconsistent with such an ideal.

G: Well, then, if no evil can befall anyone in Utopia, there will simply be no demand there for the performance of good deeds. They will, in fact, be quite impossible, and therefore not a means for gaining approval. Morality is relevant only to the extent that the ideal has not been realized, but there is no room at all for morality in the ideal itself, just as there is no room for revolution in the ideal which inspires revolutionary action.

S: What about excellence in art? We certainly admire superior artistic creators, good critics, and accomplished connoisseurs.

G: You will no doubt find what I am about to suggest very hard to accept, but it strikes me that there is no place in the ideal for any of the skills you have mentioned.

S: I must admit, Grasshopper, that I find your suggestion positively staggering. How on earth do you arrive at such a strange conclusion?

G: I believe that these skills would not exist in Utopia because art would not exist there. Art has a subject matter which consists in the actions and passions of men: with human aspirations and frustrations, hopes and fears, triumphs and tragedies, with flaws of character, moral dilemmas, joy and sorrow. But it would seem that none of these necessary ingredients of art could exist in Utopia.

S: Perhaps a good deal of art would be impossible for the Utopians, but surely not all of it. There is, or at least there used to be, a school of aesthetics which regarded art as essentially consisting in pure forms, so that content was either adventitious and therefore dispensable or, preferably, not present at all. Art as shape or design or form does not require the kind of subject matter you are talking about.

G: My own belief is that form is not separable from content in the way you suggest, but if it were, then the creation of designs, whether in tones, shapes, colours, or words could, and presumably would, be turned over to computers, since the products to be turned out would be, by hypothesis, uninspired by human emotion.

S: Even if the Utopians could not admire workers in the field of the arts, they could still admire accomplished thinkers: scientists, philosophers, and the like. Persons, that is, who are engaged in the acquisition of knowledge. Suppose we consider that possibility.

G: Very well, let us do so. Now, by hypothesis, we are supposing that our Utopians have completely eliminated the need for any instrumental activity whatever. But the acquisition of knowledge, just like the acquisition of anything else, is an instrumental process; that is, acquisition is instrumental to possession, no matter what it is that one is seeking to possess – food and shelter or knowledge. And just as we have supposed that our Utopians have acquired all the economic goods they can use, we must assume that they have acquired all the knowledge there is. In Utopia, therefore, there are no scientists, philosophers, or any other intellectual investigators.

s: Then it seems that there is nothing that one could do in Utopia in order to gain approval. But we were talking about approval only to try to discover whether such things as love and friendship could exist in Utopia. And human relationships like love and friendship include more than approval. Just as important, surely, is the *sharing* which is generally recognized to be very prominent in love and friendship. And mutual interest in something does not imply a deficiency to be over-come on the part of those who have such an interest.

G: True enough, Skepticus, but in Utopia what is there left to share? The sharing which admittedly plays a large part in love and friendship cannot be the sharing of love and friendship themselves. There must be something else; something like success and failure, adversity and prosperity, the enjoyment or creation of art, intellectual inquiry, respect for the moral qualities each possesses, etc. There is simply nothing of any importance in Utopia to be shared, so that if love and friendship could exist in Utopia, they would have to be kinds which contained neither approval nor shared interests; at most, therefore, extremely attenuated forms of love and friendship.

s: Grasshopper, let me collect my wits. In Utopia man cannot labour, he cannot administer or govern, there is no art, no morality, no science, no love, no friendship. The only thing which our analysis has not utterly destroyed is sex. Perhaps the moral ideal of man is just a supreme orgasm.

p: Dear me!

G: Of course, we mustn't forget game playing. That has not been ruled out.

s: No doubt, no doubt. Are we then to conclude that the ideal of existence is sex and games or, as we might say, fun and games?

G: Actually, now that I think of it, I am no longer all that sure about sex.

s: Oh, come now, Grasshopper!

G: No, Skepticus, I am quite serious. The obsessive popularity that sex has always enjoyed is, I suspect, inseparably bound up with man's non-Utopian condition. Sex, as we have come to know and love it, is part and parcel with repression, guilt, naughtiness, domination and submission, liberation, rebellion, sadism and masochism, romance,

and theology. But none of these things has a place in Utopia. Therefore, we ought at least to face the possibility that with the removal of
all of these constituents of sex as we value it, there will be little left
but a pleasant sensation in the loins – or wherever. People like
Norman Brown in his book *Life Against Death** take the view that
sex is something which has been distorted and corrupted by the
repressions and restraints of civilization, and that with the end of
civilization (which Brown looks forward to with great keenness), sex
will re-emerge as the unsullied item that it was in our infancies. We
will then all become happy children once again, enjoying without
inhibition our polymorphous perversity. But if, as I believe, sex is the
product rather than the victim of civilization, then when civilization
goes, sex – at least as a very highly valued item – goes as well. In
general, Skepticus, I find the current (or at least the recent) vogue
enjoyed by the injunction to 'let it all hang out' unwise in a fundamental respect. I have no quarrel with the *act* of letting it all hang
out, for that, as when we undo a very tight belt or girdle, can produce
a profound satisfaction. But once the act of permitting to hang out
whatever it is we wish to hang out has been completed, and the
attendant relief enjoyed, all we are left with in the end is just a lot of
things hanging out. And in the absence of any new constraints upon
them they just continue to hang there, a kind of pendulous monu-
ment to volitional entropy.

s: If not convinced, I am for the moment silenced.

G: Very well. Then we appear to be left with game playing as the only
remaining candidate for Utopian occupation, and therefore the only
possible remaining constituent of the ideal of existence.

s: And now I suppose you are going to rule out game playing as well.
Grasshopper, I begin to suspect that what you are really up to is to
show that the concept of Utopia itself is paradoxical, as philosophers
from time to time try to show that the alleged perfections of the Deity
entail paradoxes.

G: Quite the contrary, Skepticus. I believe that Utopia is intelligible, and
I believe that game playing is what makes Utopia intelligible. What
we have shown thus far is that there does not appear to be any thing to
do in Utopia, precisely because in Utopia all instrumental activities
have been eliminated. There is nothing to strive for precisely because

---

* Norman O. Brown *Life Against Death: The Psychoanalytical Meaning of History*
(Wesleyan University Press 1959)

everything has already been achieved. What we need, therefore, is some activity in which what is instrumental is inseparably combined with what is intrinsically valuable, and where the activity is not itself an instrument for some further end. Games meet this requirement perfectly. For in games we must have obstacles which we can strive to overcome *just so that* we can possess the activity as a whole, namely, playing the game. Game playing makes it possible to retain enough effort in Utopia to make life worth living.

s: What you are saying is that in Utopia the only thing left to do would be to play games, so that game playing turns out to be the whole of the ideal of existence.

G: So it would appear, at least at this stage of our investigation.

s: I don't think so.

G: I beg your pardon?

s: I don't think that conclusion follows.

G: You don't.

s: I believe we made a mistake earlier on.

G: A mistake.

s: Yes. Earlier on.

G: Perhaps you would be good enough to point it out to me.

s: I shall be happy to do so. When you were advancing the view that science, or any kind of intellectual inquiry, was an instrumental activity and thus could have no place in the moral ideal of man, I had some misgivings, and now I believe I know why. You know, Grasshopper, as well as I do, that people who are seriously engaged in the pursuit of knowledge value that pursuit at least as much as they do the knowledge which is its goal. Indeed, it is a commonplace that once a scientist or philosopher after great effort solves a major problem he is very let down, and far from rejoicing in the possession of his solution or discovery, he cannot wait to be engaged once more in the quest. Success is something to shoot at, not to live with. And of course, now that I think of it, this is true not only of intellectual inquiry, but it certainly can be true of any instrumental activity whatever, and frequently is. We might call this state of affairs the Alexandrian condition of man, after Alexander the Great. When there are no more worlds to conquer we are filled not with satisfaction but with despair.

G: How do you think we could have made such an elementary mistake, Skepticus?

s: I think we failed to take note of the fact that an activity which is, from one point of view, instrumentally valuable can, from another point of view, be intrinsically valuable. Thus, we would agree that carpentry is an instrumental activity; that is, instrumental to the existence of houses. But to a person who enjoys building for its own sake, that otherwise instrumental activity has intrinsic value as well. And the same could be true of anyone who really enjoys his work, whatever that work might be. It seems to follow from this that we may now re-instate most of the activities we thought we were obliged to banish from Utopia. The ideal, therefore, does not consist wholly in game playing.

G: I believe you are correct, Skepticus, in pointing out that otherwise instrumental activities can be valued as ends in themselves. But I am not convinced that it follows from that fact that game playing is not the only possible Utopian occupation. Let me see if I can persuade you of this. Let us continue to think of the moral ideal of man as an actual Utopian community, then, but where, instead of supposing that all – so to speak – *objectively* instrumental activities have been banished – physical and intellectual labour, and the like – what has been banished is simply all activity which is not *valued* intrinsically, thus leaving it open to any Utopian to enjoy the exertions of productive endeavour. Thus, just as some Utopians will be able to pluck yachts and diamonds off Utopian trees, others will be able to pluck off opportunities to fix the kitchen sink, to solve economic problems, to push forward the frontiers of scientific knowledge, and so on, with respect to anything a Utopian might find intrinsically valuable.

s: Yes, Grasshopper. That seems a much more satisfactory picture of Utopia and of the ideal of existence.

G: Splendid. Now, to continue. It is clear, I should think, that the opportunity to work – or whatever other instrumental activity it might be which is desired – should not be left to chance in Utopia. If, at any given period of time, *everyone* in Utopia wanted to work at something, then such work should be available for them all. And if nobody wanted to work, then it would not follow (as it surely would in our present non-Utopian existence) that society would collapse. And similarly, of course, with intellectual inquiry. That is to say, with respect to any objectively instrumental activity whatever, it would have to be the case that such activity *could* be undertaken, but it would also have to be the case that no such activity *need* be undertaken. For another way of saying that the Utopians only do those things which they value intrinsically is to say that they always do things because they want to, and never because they must.

s: Yes, that seems correct.

g: Very well. Now let us consider two cases that would inevitably arise in Utopia.

Case One: John Striver has spent his first decade in Utopia doing all the things that newcomers to Utopia usually do. He has travelled round the world several times, loafed a good deal in the sun, and so on, and now, having become bored, he wants some *activity* to be engaged in. He therefore makes a request (to the Computer in Charge or to God or whatever) saying that he wants to *work* at something, and he selects carpentry. Now, there is no demand for houses which John's carpentry will serve, because all the houses of whatever possible kind are already instantly available to the citizens of Utopia. What kind of house, then, should he build? Surely it would be the kind whose construction would give him the greatest satisfaction, and we may suggest that such satisfaction would require that building the house would provide enough of a challenge to make the task interesting while not being so difficult that John would utterly botch the job. Now, what I would like to put to you, Skepticus, is that this activity is essentially no different from playing golf or any other game. Just as there is no need, aside from the game of golf, to get little balls into holes in the ground, so in Utopia there is no need, aside from the activity of carpentry, for the house which is the product of that carpentry. And just as a golfer could get balls into holes much more efficiently by dropping them in with his hand, so John could *obtain* a house simply by pressing a telepathic button. But it is clear that John is no more interested in simply *having* a house than the golfer is in *having* ball-filled holes. It is the *bringing about* of these results which is important to John and to the golfer rather than the results themselves. Both, that is to say, are involved in a voluntary attempt to overcome unnecessary obstacles; both, that is to say, are playing games. This solution, it is interesting to note, was also open to Alexander the Great. Since he had run out of worlds to conquer by impetuously conquering the only world there was, he *could* have given it all back and started over again, just as one divides up the chess pieces equally after each game in order to be able to play another game. Had Alexander done that, his action would no doubt have been regarded by his contemporaries as somewhat frivolous, but from the Utopian point of view his failure to take such an obvious step would indicate that Alexander did not really place all that high a value on *the activity of conquering worlds*.

    Case Two: The early experience of William Seeker in Utopia is very similar to that of John Striver. William, too, after a time, wishes to be able to achieve something. But whereas John's abilities and interests had led him to choose a manual art, William is led to choose the pursuit of scientific truth. Now again, how much scientific inquiry there is to undertake at any given time cannot be left to chance, since the interests in doing scientific research might far exceed the amount of research that could logically be undertaken at any given time. It is even conceivable that there would come a time when all scientific investigation had come to an end; a time, that is, when everything knowable was in fact known. Since, therefore, there could be no guarantee that there would always be an objective opportunity to do scientific research, it follows that it would be undesirable to have Utopian scientists stop doing research on a problem simply because the problem had already been solved. For what is important in Utopia is not the objective state of scientific knowledge, but the *attitude* of the Utopian scientist, which may be described in the following way. If the solution of the problem he is working on were readily retrievable from the memory banks of the computers, the Utopian scientist would not retrieve the solution. This is just like the devotee of crossword puzzles who knows that the answers to the puzzle will be published next day. Still, he tries to solve the puzzle today, even though there is no urgency whatever in having the solution today rather than tomorrow. And just as the dedicated puzzle solver will say, 'Don't tell me the answer; let me work it out for myself,' William Seeker will have the same attitude towards his scientific investigations. Even if other means for coming to know the answer are readily available, he voluntarily rejects these means so that he will have something to do. But this is again, I submit, to play a game.

s: What you seem to be saying is that a Utopian could engage in all of the achieving activities that normally occupy people in the non-Utopian world, but that the quality, so to speak, of such endeavours would be quite different.

G: Yes. The difference in quality, as you put it, can be seen in the contrast in attitude of a lumberjack when he is, on the one hand, plying his trade of cutting down trees for the sawmill and, on the other hand, when he is cutting down trees in competition with other lumberjacks at the annual woodcutter's picnic. Thus, all the things we now regard as trades, indeed all instances of organized endeavour whatever, would, if they continued to exist in Utopia, be sports. So that in addition to hockey, baseball, golf, tennis, and so on, there would also be the sports of business administration, jurisprudence, philosophy, production management, motor mechanics, *ad*, for all practical purposes, *infinitum*.

s: So that the moral ideal of man does, after all, consist in game playing.

G: I think not, Skepticus. For now that the Utopians have something to do, both admiration and sharing are again possible, and so love and friendship as well. And with the re-introduction of the emotions associated with striving – the joy of victory, you know, and the bitterness of defeat – emotional content is provided for art. And perhaps morality will also be present, possibly in the form of what we now call sportsmanship. So, while game playing need not be the sole occupation of Utopia, it is the essence, the 'without which not' of Utopia. What I envisage is a culture quite different from our own in terms of its *basis*. Whereas our own culture is based on various kinds of scarcity – economic, moral, scientific, erotic – the culture of Utopia will be based on plenitude. The notable institutions of Utopia, accordingly, will not be economic, moral, scientific, and erotic instruments – as they are today – but institutions which foster sport and other games. But sports and games unthought of today; sports and games that will require for their exploitation – that is, for their mastery and enjoyment – as much energy as is expended today in serving the institutions of scarcity. It behoves us, therefore, to begin the immense work of devising these wonderful games now, for if we solve all of our problems of scarcity very soon, we may very well find ourselves with nothing to do when Utopia arrives.

s: You mean we should begin to store up games – very much like food for winter – against the possibility of an endless and endlessly boring summer. You seem to be a kind of ant after all, Grasshopper, though, I must admit, a distinctly odd kind of ant.

G: No, Skepticus, I am truly the Grasshopper; that is, an adumbration of the ideal of existence, just as the games we play in our non-Utopian lives are intimations of things to come. For even now it is games which give us something to do when there is nothing to do. We thus call games 'pastimes,' and regard them as trifling fillers of the interstices in our lives. But they are much more important than that. They are clues to the future. And their serious cultivation now is perhaps our only salvation. That, if you like, is the metaphysics of leisure time.

s: Still, Grasshopper, I find that I have a serious reservation about the

Utopia you have constructed. It sounds a grand sort of life for those who are very keen on games, but not everyone *is* keen on games. People like to be building houses, or running large corporations, or doing scientific research to some purpose, you know, not just for the hell of it.

G: The point is well taken, Skepticus. You are saying that Bobby Fischer and Phil Esposito and Howard Cosell might be very happy in paradise, but that John Striver and William Seeker are likely to find quite futile their make-believe carpentry and their make-believe science.

S: Precisely. (*Pause*) Well, Grasshopper, what answer do you have to make to this objection? (*There is another pause*) Grasshopper, are you dying again?

G: No, Skepticus.

S: What is it, then? You look quite pale.

G: Skepticus, I have just had a vision.

S: Good lord!

G: Shall I tell you about it?

S: (*Skepticus glances furtively at his wrist watch*) Yes. Well. Certainly, Grasshopper, please proceed.

G: The vision was evidently triggered by your suggestion that not every-one likes to play games, and it was a vision of the downfall of Utopia, a vision of paradise lost. I saw time passing in Utopia, and I saw the Strivers and the Seekers coming to the conclusion that if their lives were merely games, then those lives were scarcely worth living. Thus motivated, they began to delude themselves into believing that man-made houses were more valuable than computer-produced houses, and that long-solved scientific problems needed resolving. They then began to persuade others of the truth of these opinions and even went so far as to represent the computers as the enemies of mankind. Finally they enacted legislation proscribing their use. Then more time passed, and it seemed to everyone that the carpentry game and the science game were not games at all, but vitally necessary tasks which had to be performed in order for mankind to survive. Thus, although all of the apparently productive activities of man were games, they were not believed to be games. Games were once again relegated to the role of mere pastimes useful for bridging the gaps in our serious endeavours. And if it had been possible to convince these people that they were in fact playing games, they would have felt that their whole lives had been as nothing – a mere stage play or empty dream.

S: Yes, Grasshopper, they would believe themselves to be nothing at all, and one can imagine them, out of chagrin and mortification, simply vanishing on the spot, as though they had never been.

G: Quite so, Skepticus. As you are quick to see, my vision has solved the final mystery of my dream. The message of the dream now seems perfectly clear. The dream was saying to me, 'Come now, Grasshopper, you know very well that most people will not want to spend their lives playing games. Life for most people will not be worth living if they cannot believe that they are doing *something* useful, whether it is providing for their families or formulating a theory of relativity.'

S: Yes, it seems a perfectly straightforward case of an anxiety dream. You were acting out in a disguised way certain hidden fears you had about your thesis concerning the ideal of existence.

G: No doubt. But tell me, Skepticus, were my repressed fears about the fate of mankind, or were they about the cogency of my thesis? Clearly they could not have been about both. For if my fears about the fate of mankind are justified, then I need not fear that my thesis is faulty, since it is that thesis which justifies those fears. And if my thesis is faulty, then I need not fear for mankind, since that fear stems from the cogency of my thesis.

S: Then tell me which you feared, Grasshopper. You alone are in a position to know.

G: I wish there were time, Skepticus, but again I feel the chill of death. Goodbye.

S: Not goodbye, Grasshopper, *au revoir*.